I0057429

Ethnic Healing Herbs for COLD, FLU AND LUNGS AILMENTS

Ethnic Healing Herbs for COLD, FLU AND LUNGS AILMENTS

S.K.Sood
Shefali Kaushal
T.N. Lakhanpal
Suresh Kumar
Department of Biosciences
Himachal Pradesh University
Shimla – 171 005

2011
DAYA PUBLISHING HOUSE®
Delhi - 110 035

© 2011 AUTHORS
ISBN 9789351240372

All rights reserved, including the right to translate or to reproduce this book or parts thereof except for brief quotations in critical reviews.

Published by	:	**Daya Publishing House®** **A Division of** **Astral International Pvt. Ltd.** **– ISO 9001:2008 Certified Company –** 4760-61/23, Ansari Road, Darya Ganj New Delhi-110 002 Ph. 011-43549197, 23278134 E-mail: info@astralint.com Website: www.astralint.com
Laser Typesetting	:	**Classic Computer Services** Delhi - 110 035
Printed at	:	**Chawla Offset Printers** Delhi - 110 052

PRINTED IN INDIA

Preface

The dependence of modern healthcare systems upon plant resources is increasing day by day. Therefore, the demand on phytotherapeutics has increased many folds in both developing and developed countries due to the growing recognition that these are safe, cheap, and easily available and with no side effects. Unlike India, traditional medicine is an integral part of the formal healthcare system in China and is being utilised in about 40 per cent of cases at the primary care level. Even the 'World Health Organisation' is advocating its use in meeting primary healthcare needs to achieve the goal of 'health for all'. Despite innumerable advantages of plant-based therapeutics, no efforts have been made till to-date to put together widely scattered knowledge on plants used for curing various lungs ailments from myriad sources. Based on this rationale, the present compendium was conceived with all the earnest efforts to provide base line data for discernible users. This may go a long way in integration of indigenous system of medicine with modern medicine for alleviating human suffering. The authors believe that this volume will provide overall glimpse of the subject and should therefore, find a place in shelves of libraries and scholars as a very useful reference source.

This venture could not have been possible without the valuable advice and incessant encouragement of Prof. S.K. Gupta, Vice-Chancellor, Himachal Pradesh University, Shimla, for which we express our sincere gratitude to him. Well deserved appreciation is due to Prof. D.C. Kalia and Dr V.K. Mattu and other colleagues of Department of Biosciences for their co-operation and well wishes; to Mr Dhiraj Rawat, Ms Anjana and Mr Munish for their ungrudging help in critically reading the various drafts of this manuscript, and to Mr Bal Krishan for his help in various ways. Thanks are also due to authors of books, papers and figures listed in the compendium. Finally, we would like to thank the members of our families for moral support and unflinching patience during the writing of this manuscript. We are also thankful to the publisher for publishing the book in its present form.

S.K.Sood

Shefali Kaushal

T.N. Lakhanpal

Suresh Kumar

Contents

Abbreviations

Δ	Delta		m	Meter
Ψ	Psi		M.P.	Madhya Pradesh
α	Alpha		Mg	Magnesium
β	Beta		mg	Milligram
γ	Gamma		N	North/Northern
ω	Omega		N.E.	North-East
alc.	Alcoholic		N.W.	North-West
alt.	Altitude		OH	Hydroxy
aq.	Aqueous		Pl.	Plate
C	Central		S	South/Southern
E.	East/Eastern		S.E.	South-East
Eng.	English		S.W.	South-West
et al.	et alia (and other authors)		Sans.	Sanskrit
etc.	et cetera (so on)		sp	Species (singluar)
EtOH	Ethanol		spp	Species
ext.	Extract		Syn.	Synonym
G	Gram		var.	Variety
H.	Hindi		Vit.	Vitamin
H.P.	Himachal Pradesh		W.	West/Western
i.e.	id est (that is)		WHO	World Health Organisation

Introduction

Plant kingdom was man's earliest source for vital medicines and estimated to date back to 4,000 B.C. *Atharvaveda* has been considered the first documented work dealing with the medicinal plants of India. Following the advent of modern system of medicine, herbal medicines suffered a setback, but in recent years, indigenous knowledge about medicine has gained importance all over the world as plant based derivatives are considered cheap, safe and effective than modern medicines. About 80 per cent of the world's population depends wholly or partially on traditional medicine for its primary healthcare needs (Wambehe 1990). Of the 119 plant based drugs being used today about 74 per cent are from traditional medicine (Sinha and Sinha 2002). The Indian subcontinent is ethnically, floristically and agriculturally very diverse (Arora 1995), and our country happens to be the second largest in the world in respect of human population. Nowadays, the folk medicine mainly based on plant enjoys a respectable position in developing countries, where modern health service is limited (Rajasekharan *et al.,* 1996). In this context, it is pertinent to mention that out of the 15,000 species of flowering plants occurring in India, 1,500 plant species are being used in Indian system of medicine, i.e Ayurveda (Bhatnagar 1997) from which plant drugs worth Rs 3,400 million per annum are utilized for its various preparations (Gupta 1986). Farnsworth (1985)- Director of WHO Collaborating Centre for Traditional Medicine strongly urged the need to scientifically investigate all those plants which are used by the primitive folk-healers and ethnic societies of world with the hope to isolate the active principles from them against various dreaded ailments.

Cold, Flu and Lungs Ailments:
Current Status and Perspective

India is also rich in traditional knowledge as evidenced by the occurrence of over 550 native or tribal communities constituting about 7.7 per cent of the total population of the country, covering more than 5,000 villages (Anonymous 1994; Saklani and Jain 1994). The pioneer work for systematic record of medicinal plants of India goes to Garcia- de- Orta, Physician and Governor of Goa (1563). Afterwards, a lot of studies have been carried out on economic and medicinal plants of India, and these have been amply reviewed by Jain *et al.* (1984), Jain and Srivastava (2001) and Jain (2002). It is a well known fact that the last few decades have witnessed an upsurge of activities in documenting ethnic knowledge regarding medicinal plants. Recently, in his analysis of about 1600 references presented in a recent detailed bibliographic work on Indian ethnobotany, Jain (2002) urged the need to carry out critical studies on diseases- specific or endemic recipies or cures among certain populations, or any particular disease and the way the local folk attempt to manage them. Unfortunately, a very little work has been done in this direction in India as the contributions on plants used in a particular disease are few and far from satisfactory, *viz.*, diabetes (Baily and Day 1989, Satyavati *et al.*, 1989, Grover *et al.*, 2002, Desai and Jasray 2003, Subramaniam and Basu 2003), ethnodermatology (Khan and Chughatai 1982), ethnogynaecology (Tarafder 1983a, b, 1984), eye diseases (Pal 1973), infectious diseases (Saxena and Vyas 1981), jaundice (Goel and Bhattacharya 1981), leprosy (Das and Sharma 1958-1959), snake-bite and ethno-orthopaediatrics (Jain 1964, 1967). So far, no one has ever attempted to present a comprehensive account of the plants used by the traditional Indian communities for the cure of various lungs ailments, such as asthma, chronic bronchitis, pneumonia, tuberculosis, lungs cancer, etc. which as a group affects a large number of people and cause of death in our country. The present work was, therefore, undertaken with the view-to document information in this aspect available from the Indian perspective. It is hoped that this treatise will fill the gap adequately and will be of immense use in integrating traditional knowledge with modern system of medicine for the general welfare of human.

Enumeration

The present compilation on the ethnic plants used for the cure of cold, flu and lungs ailments is the outcome of literary research based on data compiled and collected from classical literature, other books and recent articles published in standard reference journals which are not easily available to educated layman, *viz.* (i) *Indigenous Drugs of India* (Dey 1896), (ii) *Indian Medicinal Plants* (Kirtikar and Basu 1984), (iii) *Flora Indica or Description of Indian Medicinal Plants* (Roxburgh 1932), (iv) *The Wealth of India: A Dictionary of Indian Raw Materials and Industrial Products* (Anonymous 1948-1976), (v) *Glossary of Indian Medicinal Plants* (Chopra *et al.*, 1956), (vi) *Supplement to the Glossary of Indian Medicinal Plants* (Chopra *et al.*, 1969), (vii) *Dictionary of Economic Plants in India* (Singh *et al.*, 1983), (viii) *The Useful Plants of India* (Ambasta 1986), (ix) *Economic Plants of India* (Agarwal 1986), (x) *Medicinal Plants of India* (Satyavati *et al.*, 1987), (xi) *Economic Plants of India* (Nayar *et al.*, 1989), (xii) *A Dictionary of Economic Products* (Watt 1972), (xiii) *Fruit Drug Plants of India* (Das and Agarwal 1991), (xiv) *The Treatise on Indian Medicinal Plants* (Chatterjee and Pakrashi 1991-1995), (xv) *Indian Medicinal Plants- A Compendium of 500 Species* (Warrier *et al.*, 1994-96), (xvi) *Indian Materia Medica* (Nadkarni 1995), (xvii) *Ethnobotany in South Asia* (Maheshwari 1996), (xviii) *Handbook of Medicinal Plants* (Bhattacharjee 1998), (xix) *Data Base on Medicinal*

Plants Used in Ayurveda (Sharma *et al.*, 2001), (xx) *Encyclopedia of Indian Medicinal Plants* (Khare 2004), etc. so as to provide information on inexpensive phytotherapeutic agents against the cure of cold, flu and lungs ailments affecting mankind. For simplicity and convenience of reference, the information is supplemented with their valid botanical names and synonyms, family, *Eng., Sans. and H. Names,* distribution, part/s used, active constituents and the relevant references, wherever known. As many as seven illustrations encompassing 20 color photographs have been included to enable a lay reader to identify some of the plants used for curing cold, flu and lungs ailments.

Appendices include information on index to total number of genera and species of plants under various divisions of plant kingdom for cure of cold, flu and lungs ailments, index to families, index to various species under a genus, index to botanical names, index to English names, index to Sanskrit names, index to Hindi names and glossary to some important medical terms and their meanings.

Ethnic Healing Herbs for Cold, Flu and Lungs Ailments

Abelmoschus esculentus (L.) Moench.

Syn.: *Hibiscus esculentus* L.

Family: Malvaceae

Hindi Name	:	H: Bhindi.
Distribution	:	Cultivated throughout India.
Part/s Used	:	Roots and Fruits.
Utilization	:	5g powdered roots given with cow milk, twice a day for a month, to cure tuberculosis. Fruits boiled in milk prescribed for cough. Vapours from a hot decoction of fruit effective in irritable condition of the throat and cough of phthisis.
Active Constituents	:	Fruit contains glucose, glucuronic acid, galacturonic acid, quercetin and hyperin.
References	:	Bennet (1987), Das *et al.* (1991), Khanna *et al.* (1996), Rastogi and Mehrotra (1990, 1995).

Abelmoschus ficulneus (L.) Wt. and Arn. *ex* Wt.

Syn.: *Hibiscus ficulneus* Linn.

Family: Malvaceae

Distribution	:	Punjab to Bengal; A.P to S. India.
Part/s Used	:	Seeds.
Utilization	:	Seeds considered good against asthma.
Active Constituents	:	Seed oil rich in epoxyoleic (4.9 per cent), malvalic (4.2 per cent) and sterculic (1.0 per cent) acids.
References	:	Rastogi and Mehrotra (1993), Singh and Pandey (1980).

Abies pindrow Royle

Syn.: *A. pindrow* Spach.; *A. webbiana* Lindle.
var. *pindrow* Brandis; *Pinus pindrow* Royle

Family: Pinaceae

Eng. and H. Names	:	Eng. : Silver fir, West Himalaya low level fir.
		H: Morinda, Ragha.

Distribution	:	Temperate Himalaya from Kashmir to Nepal: 2,100-3,600m.
Part/s Used	:	Leaves.
Utilization	:	1tsp mixture of paste of fresh leaves and honey in equal quantity, given twice a day, for cough and cold.
Active Constituents	:	Leaf oil contains α-pinene, l- limonene, D^3-carene, dipentene, l-cadinene and l-bornyl-O.
Biological Activity	:	50 per cent extract of aerial parts spasmolytic.
References	:	Asolkar *et al.* (1992), Pandey and Pandey (1999).

Abrus precatorius Linn.

Family: Fabaceae

Eng., Sans. and H. Names	:	Eng.: Indian liquorice, Jequirity, Rosary pea;
		Sans: Aruna, Gunja, Gunjika, Tamrika, Vanya;
		H: Gunchi, Gunja, Rati.
Distribution	:	All along the Himalaya ascending to 1,000m and spreading through the plains of India.
Part/s Used	:	Leaves, Roots and Seeds.
Utilization	:	Powdered, roasted leaves taken with honey twice a day for irritating cough and throat problem. Extract of root given for chronic cough. Dried root powder (5-10 g) in combination with honey given for a week to cure cough and cold. 5-10 ml decoction of roots in combination with rhizome of ginger and *Acorus calamus* Linn. good against cough. Crushed seeds also found to check pneumonia and other bronchial diseases.
Active Constituents	:	These are abrin, abrine 7 abarnin, alanine, serine and valine, abrusin and its 2″-O-apioside (seeds). Precol, abrol and 2 alkaloids- abraisine and precascine (roots).
Biological Activity	:	Saponins containing oleanolic acid have been found to exhibit strong pharmacological activity.
References	:	Ambasta (1986), Chaudhary and Hutke (2002), Jadeja (1999), Kirtikar and Basu (1984), Painuli and Maheshwari (1996), Prasad *et al.* (1996), Rastogi and Mehrotra (1990, 1991, 1993, 1995), Sharma (2003), Singh *et al.* (1996), Watt (1972).

Abutilon indicum (Linn.) Sw.

Syn.: *Sida indica* Linn.; *Sida guineensis* Schumach.;
***A. asiaticum* (L.) Sweet; *A. indicum* G. Don**

Family: Malvaceae

Eng., Sans. and H. Names	:	*Eng.*: Country mallow, Indian Mallow;
		Sans.: Ghanta, Kankatika, Shita, Vikankata;
		H.: Kanghi, Kandhi, Tepari.
Distribution	:	Throughout tropics upto 600m.
Part/s Used	:	Leaves, Roots and Seeds.
Utilization	:	Decoction of leaves and that of seeds prescribed for bronchitis. About 10g powdered roots given with water, twice a day for 3 months to cure tuberculosis.
Active Constituents	:	Main ingredients are raffinose (1.6 per cent) and semi drying oil (9.21 per cent) consisting of linoleic, oleic, palmitic and stearic acid (seeds), glucose, fructose, galactose, tocopherol oil (0.3 per cent) (leaves).
References	:	Bennet (1987), Bhalla *et al.* (1982), Caius (2003), Chopra *et al.* (1980), Khanna *et al.* (1996), Kirtikar and Basu (1984), Rastogi and Mehrotra (1993), Shah *et al.* (1983), Singh and Pandey (1980).

Acacia arabica Willd.

Syn.: *A. nilotica* (Lam.) Willd; *Mimosa arabica* Lamk.; *Acacia vera* Willd.

Family: Fabaceae

Eng., Sans. and H. Names	:	*Eng.*: Black Babool, Indian gum Arabic tree;
		Sans.: Ajabaksha, Babbula, Barbura;
		H.: Babla, Babul, Babur.
Distribution	:	Cultivated throughout the greater part of India.
Part/s Used	:	Leaves, Pods, Bark and Gum.
Utilization	:	'Laddus' containing fried gum taken to cure cough. Also, gum with honey (10g) and latex (3 g) of *Calotropis procera* are taken orally by 'Bhils' thrice a day against asthma. Decoction of 50g bark crushed with a little jaggery and taken orally for 4 days in whooping cough, asthma and bronchitis. Leaves and pods also useful in bronchitis and cough.
Active Constituents	:	Chief constituents are quercitin, catechol, gallic acid, (+) epicatechin, (-) epigallocatechin-5, 7-digallate and tannins (pods, bark), polyphenols like catechin,

epicatechin, dicatechin, quercetin and tannin, arabinobiose–2–O–β–L–arabinofuranosyl–L–arabinose along with known 3–O–β–L–arabinopyranosyl–L–arabinose (gum).

References	:	Das and Agarwal (1991), Katewa *et al.* (2001), Kirtikar and Basu (1984), Rastogi and Mehrotra (1990) Sen and Batra (1997), Sharma (2003), Singh and Pandey (1980, 1998), Thakur *et al.* (1989), Tripathi and Kumar (2003), Watt (1972).

Acacia catechu Willd.

Syn.: *Mimosa catechuoids* Roxb.; *Acacia catechuoids* Wall.; *A. polycantha* Willd.; *A. sundra* Willd.

Family: Fabaceae

Eng., Sans. and H. Names	:	*Eng.*: Black catechu, Catechu nigrum, Cutch tree;
		Sans.: Bahushalya, Balapatra, Karkali, Khadira;
		H.: Katha, Khair, Khairbabul.
Distribution	:	Sub-Himalayan tracts, Semi–arid areas of C. and S. India.
Part/s Used	:	Stem-bark.
Utilization	:	2-3 tsp decoction of stem without bark given twice a day, for treating cough. Decoction of Stem-bark with that of *Acacia ferruginea, Holarrhena pubescens* and roots of *Calotropis procera* taken orally at bed time by tribals of Udaipur to cure asthma and pneumonia. Also, pills prepared in combination with catechu, black pepper, clove and *Terminalia bellerica* fruits taken orally to cure cough.
Active Constituents	:	Chief constituents are catechin, catechutanic acid and tannin.
Biological Activity	:	Et.OH extract of stem spasmolytic.
References	:	Ambasta (1986), Asolkar *et al.* (1992) Kirtikar and Basu (1984), Panthi and Chaudhary (2003), Sharma (2003), Singh and Pandey (1998), Thakur *et al.* (1989).

Acacia leucophloea (Roxb.) Willd.

Syn.: *Mimosa leucophloea* Roxb.

Family: Fabaceae

Eng., Sans. and H. Names:		*Eng.*: White babool;
		Sans.: Arimela, Pitaka, Shvetarbura;
		H.: Nimbar, Panharya, Safed kikar.

Distribution	:	Punjab, C. India, Deccan.
Part/s Used	:	Bark.
Utilization	:	Bark decoction effective for treating cough, cold and bronchitis.
Active Constituents	:	Stem-bark yields isokannin, cyanin and leucodelphinidine-3-O-α-L-rhamnopyranoside, n-hexacosanol, β-amyrin and β-sitosterol.
References	:	Ambasta (1986), Kirtikar and Basu (1984), Kshirsagar *et al.* (2003), Rastogi and Mehrotra (1991, 1993), Watt (1972).

Acacia senegal Willd.

**Syn.: *A. rupestris* Stocks; *A.verek* Guill. *et.* Perrot.;
Mimosa Senegal Linn.; *M.senegalensis* Lam.**

Family: Fabaceae

Eng. and Sans. Names:		*Eng.*: Gum Arabic tree;
		Sans.: Sveta khadira.
Distribution	:	Punjab.
Part/s Used	:	Gum.
Utilization	:	Gum soaked in water given with milk for cough.
Active Constituents	:	Constituents of gum are sugars-d-galactose, L-arabinose, L–rhamose, d–glucuronic acid, toxalbumin, glucosidal acid and arabic acid.
Biological Activity	:	Et. OH extract of Stem-bark exhibits spasmolytic
References	:	Asolkar *et al.* (1992), Gill and Nyawuame (1994), Greenish (1999), Kirtikar and Basu (1984), Watt (1972).

Acalypha indica Linn.

Syn.: *A. spicata* Forsk.; *A. ciliata* Wall; *A. canescens* Wall.

Family: Euphorbiaceae

Eng., Sans. and H. Names	:	*Eng.*: Indian Acalypha;
		Sans.: Harita manjari;
		H.: Khokali, Kuppi.
Distribution	:	Almost throughout India.
Part/s Used	:	Whole Plant.
Utilization	:	10ml juice of the plant taken twice a day mixed with honey for asthma and bronchial infection. 1 tsp extract of leaves given twice a day against asthma, bronchitis, cough and pneumonia. Root juice or root powder taken

with milk removes the catarrhal matter and phleg from bronchial tubes and cures cold and pharyngitis. In homeopathy juice of herb is employed against severe cough associated with bleeding from lungs.

Active Constituents	:	These are kaempferol, acalyphine (plant), stigmasterol (root, leaves) and acalyphol acetate (leaves).
References	:	Ambasta (1986), Caius (2003), Das (1997), Jain (1968), Kshirsagar *et al.* (2003), Maya *et al.* (2003), Raju (2000), Rastogi and Mehrotra (1995), Singh and Pandey (1980), Sinha and Sinha (2001), Trivedi (2002), Watt (1972).

Acanthus ilicifolius Linn.

Syn.: *Dilivaria ilicifolius* Juss.

Family: Acanthaceae

Eng., Sans. and H. Names	:	*Eng.*: Sea holly;
		Sans.: Harikusa;
		H.: Hargoza, Harkuchkanta.
Distribution	:	Sea-coast of India.
Part/s Used	:	Roots.
Utilization	:	Roots found to be useful in asthma.
Active Constituents	:	Roots contain acanthicifoline, benzoxazolin–2 -one, octasanol, stigasterol and benzoxazolin -2-one.
References	:	Ambasta (1986), Kirtikar and Basu (1984), Rastogi and Mehrotra (1993, 1995).

Achillea millefolium Linn.

Syn.: *A. lanulosa* Nutt.

Family: Asteraceae

Eng. and H. Names	:	*Eng.*: Arrow root, Carpenter grass, Devil's nettle;
		H.: Gandana, Rojmari.
Distribution	:	Grows wild at some places of W. Himalaya between 2,000-3,000m.
Part/s Used	:	Flower heads.
Utilization	:	Powdered dried flower heads prescribed for cold.
Active Constituents	:	Inflorescence yields volatile oil (0.25 per cent) having azulene (1.5 per cent), achilloic acid, alkaoil-achi-lleine, bitter principle- ivain and tannin.

References	:	Chandler and Hooper (2003), Jain *et al.* (1997), Rastogi and Mehrotra (1990), Sharma (2003), Sharma and Sood (1997), Thakur *et al.* (1989),

Achyranthes aspera Linn.

Family: Amaranthaceae

Eng., Sans. and H. Names	:	*Eng.*: Prickly chaff flower;
		Sans.: Mayura, Mayuraka;
		H.: Latjeera.
Distribution	:	All over India, ascending upto 1,000m.
Part/s Used	:	Seeds, Roots and Entire Plant.
Utilization	:	Plant ash taken with small quantity of sugar to cure cough. A pinch of root powder along with powdered fruits of *Piper nigrum* L. given with honey to cure asthma. Two pea-sized pills made by frying 10g fresh pounded roots with 100g onion in 200ml mustard oil given, twice a day for 15 days, to cure tuberculosis. 2g mixture of powdered leaves and roots in combination with 4-5 fruits of black pepper considered good against asthma. Tribals of Udaipur chew the roots to cure asthma. Powdered roasted seeds or its curry taken orally with warm water to cure asthma. 500mg powder of ripe seeds kept mixed with latex of *Calotropis procera* for 2-3 days given, thrice daily for 3 days, against flu and whooping cough.
Active Constituents	:	These are α-L–rhamnopyranosyl (1$\rightarrow$4) β-D-glucopyranosyl (1$\rightarrow$4) β- D-glucopyranosyl (1$\rightarrow$3) oleanolic acid and β-D-galactopyranosyl (1$\rightarrow$28) ester of saponin (seeds), betaine, achyranthine and ecdysone (roots).
Biological Activity	:	Achyranthine has been found to increase the rate and amplitude of respiration in experimental animals (dogs, frogs).
Remarks	:	Ash of root considered expectorant.
References	:	Jain and Singh (1997), Katewa *et al.* (2001), Lal *et al.* (1996), Punjani (2002), Rastogi and Mehrotra (1991), Shome *et al.* (1996), Singh (2002), Singh and Pandey (1980), Thakur *et al.* (1989), Watt (1972).

Aconitum violaceum Jacq. ex Stapf.

Syn.: *A. multifidium* Royle

Family: Ranunculaceae

Distribution	:	Sub–alpine and Alpine meadows of cold arid region (Kashmir to Uttaranchal Himalaya).
Part/s Used	:	Flowers, Tubers and Roots
Utilization	:	Boiled dried roots given in small doses to cure cough. Decoction of 1-2 tubers, taken twice a day for curing cough and cold. 2-4g fermented mass of fresh flowers and sugar kept in earthen pots for 2-3 days, given twice daily for 7 days, to cure asthma and bronchitis.
Active Constituents	:	Root contains–indaconitine and bikhaconitin.
References	:	Asolkar *et al.* (1992), Chandrasekar and Srivastava (2003), Sharma (2003), Sharma and Rana (1999), Singh (1996).

Acorus calamus Linn.

Family: Araceae

Eng., Sans. and H. Names	:	*Eng.*: Calamus, Sweet flag;
		Sans.: Badhaniya, Bhadra, Golomi, Jalaja;
		H.: Bhoja, Ghorbach, Gorbach, Vacha.
Distribution	:	Grows wild in marshy places upto 2,000m in the Himalaya, Manipur and Naga hills. Some parts of S. India.
Part/s Used	:	Roots and Rhizome.
Utilization	:	Dried roots and stem chewed to treat cough and cold. Roasted rhizome (1cm long) chewed, 3 times a day, for cough, cold and flu. Paste of rhizome also given orally with milk 3 times a day, for 3-4 days against cough and cold. Due to its essential oil contents, it acts as expectorant, and promotes flow of bronchial secretions.
Active Constituents	:	Rhizome yields coradin together with 2, 4, 5 - trimethoxybenzaldehyde, 2, 5 -dimethoxybenzo- quinone, galangin, sitosterol, myristic (1.3 per cent), palmitic (18.2 per cent), palmitoleic (16.4), stearic (7.3), oleic (29.1 per cent), linoleic (24.5) and arachidic acids (3.2 per cent).
References	:	Caius (2003), Greenish (1999), Kumar and Jain (1998), Lal *et al.* (1996), Panthi and Chaudhary (2003), Rastogi and Mehrotra (1993) Saklani and Jain (1994), Sharma and Sood (1997), Singh (1999), Thakur *et al.* (1989).

Actiniopteris radiata (Swartz.) Link.

Syn.: *A. australis* (Linn. f) Link.

Family: Actiniopteridaceae

Eng. Name	:	*Eng.*: Peacocks tail.
Distribution	:	Found throughout the Aravalli hills: 400- 1,000m.
Part/s Used	:	Whole Plant and Leaves.
Utilization	:	Decoction of dry leaves cures tuberculosis. Whole plant considered good against bronchitis.
Active Constituents	:	Aerial plant parts yield hentriacontane, hentriacontanol, β- sitosterol, its palmitate and glucoside.
References	:	Bhattacharjee (1998), Rastogi and Mehrotra (1991), Sharma (2003), Sharma and Vyas (1985), Singh and Pandey (1980).

Adansonia digitata Linn.

Syn.: *Baobaus digitata* O. Ktze.

Family: Bombacaceae

Eng., Sans. and H. Names	:	*Eng.*: Baobab, Monkey bread tree, Sour gourd;
		Sans.: Chitrala, Choramli, Gopali, Gorakshi;
		H.: Gorakamali, Gorakh imli, Goramlichora.
Distribution	:	Grown in many places in India.
Part/s Used	:	Fruit Pulp.
Utilization	:	Acidic fruit pulp mixed with figs given to cure asthma in Konkan region.
Active Constituents	:	Fruit pulp rich in ascorbic acid, pectin, iron and calcium.
References	:	Ambasta (1986), Das and Agarwal (1991), Kirtikar and Basu (1984), Rastogi and Mehrotra (1993).

Adhatoda vasica Nees
[Pl. 1A]

Syn.: *Adhatoda zeylanica* Medic.; *Justicia adhatoda* L.

Family: Acanthaceae

Sans. and H. Names	:	*Sans.*: Amalaka, Bashika, Panchamukhi, Vasa;
		H.: Adalsa, Arusha, Bansa, Rusa, Vasaka.
Distribution	:	Throughout India, often cultivated.
Part/s Used	:	Leaves.

Utilization	:	Dried inflorescence burnt and inhaled for asthma by Jatapus and Savaras. In Bengal and upper India, the leaves are smoked as cheroots for curing asthma. Fresh or dried leaves constitute the drug Vasaka, used in bronchial troubles. Juice of leaves, 2-3 tsp employed for cough, bronchitis, asthma and tuberculosis. Juice of leaves and roots prescribed in chronic bronchitis, cough and asthma. An infusion of leaves and twigs given orally for cough and flu. Also, 2 -3 tsp juice of fresh leaves mixed with equal volume of honey and ginger juice useful against cough.
Active Constituent	:	Chief principle is Vasicine (0.54-1.1 per cent).
References	:	Ambasta (1986), Bhatt and Mitaliya (1999), Kirtikar and Basu (1984), Kumar (2002), Pandey *et al.* (2003), Raju (2000), Rao and Henry (1996), Singh *et al.* (2003).

Adiantum capillus-veneris Linn.

Family: Polypodiaceae

Eng. and H. Names	:	*Eng.*: Maiden hair fern, Maria's fern, Our lady's hair; *H.*: Hansraj, Mubaraka.
Distribution	:	W. Himalaya, ascending upto 700m.
Part/s Used	:	Leaves.
Utilization	:	Decoction of 2-3 fresh or shade dried leaves taken with 1 tsp honey taken orally for acute bronchitis.
Active Constituents	:	Leaves rich in β-sitosterol, stigmasterol and campesterol from leaves.
Remarks	:	In France, large quantities are employed in preparing Siropde Cappillaire, a medicine in cough and bronchitis affection.
References	:	Bhattacharjee (2001), Caius (2003), Jain *et al.* (1995), Kaul (1997), Rastogi and Mehrotra (1990, 1995), Singh and Pandey (1980).

Adiantum caudatum Linn.

Family: Polypodiaceae

Sans. Name	:	*Sans.*: Mayurashikha.
Distribution	:	Throughout India; plains and lower slopes of hills.
Part/s Used	:	Whole Plant (asthma, bronchitis), Leaf (cough).

Active Constituents	:	Chief constituents of plant are a triterpene alcohol–29–norhopan- 22–ol, β–sitosterol, its glucoside, quercetin-3-O-glucoside, filic–3–ene, 6–hentriacontanone, hentriacontane, isoadiantone, triterpenes- fernene, adiantone and isoadiantone.
References	:	Islam (1983), Jain (1984), Kirtikar and Basu (1984), Rastogi and Mehrotra (1991, 1995).

Adiantum incisum Forsk.

Family: Polypodiaceae

Part/s Used	:	Fronds.
Utilization	:	Tender fronds and its infusion good against bronchial diseases. Tender fronds eaten to cure cough.
Active Constituents	:	Hentriacontane, 16- hentriacontanone, adiantone, isoadntone, β- sitosterol and fernene isolated.
References	:	Asolkar *et al*. (1992), Nwosu (2001), Singh and Pandey (1980).

Adiantum lunnulatum Burm.

Family: Polypodiaceae

Eng., Sans. and H. Names	:	*Eng.*: Maiden hair;
		Sans.: Brachmadani, Chitrapada, Godhangri;
		H.: Hansapadi, Kalij hamp, Kalij hant.
Distribution	:	Throughout N.India in moist places.
Part/s Used	:	Leaves.
Utilization	:	Decoction of leaves prescribed for cough and asthma.
References	:	Bhattacharjee (1998), Chopra *et al*. (1956), Kirtikar and Basu (1984), Singh (1973), Singh and Pandey (1998), Sharma and Vyas (1985).

Adiantum pedatum Linn.

Family: Polypodiaceae

Eng. Name	:	*Eng.*: America maiden hair fern.
Distribution	:	N.W. Himalaya–Kashmir to Sikkim: 2,000-3,000m.
Part/s Used	:	Rhizome.
Utilization	:	Decoction of rhizome used as an expectorant for treating cough and cold.

Active Constituents	:	Rhizome possesses fernene, isofernene, filicene, filicinal, adiantone, adipedatol, sterols, fatty acid, volatile oil, tannins, p- coumaric, caffeic, ferullic and vanillic acid.
References	:	Chopra *et al.* (1956), Singh (1999).

Aerva lanata Juss. *ex* Schult.

Syn.: *A. floribunda* Wight.

Family: Amaranthaceae

Sans. and H. Names	:	*Sans.*: Astmabayda;
		H.: Gorkhabundi, Kapurijadi.
Distribution	:	Throughout the plains of India ascending to 1,000m in the hills
Part/s Used	:	Whole Plant.
Utilization	:	Decoction of whole plant used to cure pneumonia.
Active Constituents	:	Plant contains 4 flavonoid glycosides (I, II, III, IV), β-sitosterol, its glucoside, α-amyrin, betulin, campesterol, chrysin, daucosterol, syringic acid, vanilic acid, feruloyltyramine, feruloylhomovanillylamine, 4 alkaloids- aervine, methyl-aervine, aervoside and aervolanine.
References	:	Ambasta (1986), Chopra *et al.* (1956), Kirtikar and Basu (1984), Rastogi and Mehrotra (1995), Singh and Pandey (1980).

Agaricus igniarius (L.) E.H.L. Krause

Family: Agaricaceae

H. Names	:	*H.*: Garigond, Gharikum.
Part/s Used	:	Fruiting Body.
Utilization	:	Powdered dried fruiting body in combination with liquorice employed against chronic bronchitis and asthma.
References	:	Kitikar and Basu (1984), Watt (1972).

Ageratum conyzoides Linn.

Family: Asteraceae

Part/s Used	:	Flower heads (asthma).
Reference	:	Jain *et al.* (1997).

Ailanthus altissima (Mill.) Swingle

Syn.: *A. glandulosa* Desf.; *Toxicodendron altissimum* Mill.

Family: Simaroubaceae

Eng. Names	:	*Eng.*: Chinese Sumach, Tree of the gods, Tree of heaven.
Distribution	:	Cultivated in the hills of Punjab.
Part/s Used	:	Bark.
Utilization	:	Used for treating asthma.
Active Constituents	:	Stem-bark yields fatty acids (27 per cent), β-sitosterol, scopoletin, indole alkaloids and bitter quassinoid-shinjudilactone.
References	:	Ambasta (1986), Bhattacharjee (1998), Kirtikar and Basu (1984), Rastogi and Mehrotra (1990, 1993, 1991).

Ailanthus excelsa Roxb.

Syn.: *Pongelion wightii*. Van Tiegh

Family: Simaroubaceae

Eng., Sans. and H. Names	:	*Eng.*: Tree of heaven;
		Sans.: Aralu, Madala, Maharukha;
		H.: Limbado, Maharuk, Maharukha.
Distribution	:	Indian Peninsula, often planted in various parts of India.
Part/s Used	:	Bark and Leaves.
Utilization	:	Inhalation of vapours of stem-bark boiled in water considered good by tribals for cough and cold. Also, juice of leaves and decoction of bark given for chronic asthma and bronchitis.
Active Constituents	:	Bark contanis triacontane, hesatriacontane, a non glycosidal bitter excelsin, 13 (18) -dehydroexcelsin, glaucarubol, a triterpene, formyl-ailanthinol, β-sitosterol, vitexin and melanthine (0.09 per cent).
References	:	Ambasta (1986), Bennet (1987), Bhatt *et al.* (2002), Bhattachajee (1998), Joshi (1982), Kirtikar and Basu (1984), Kothari and Londhe (1999), Mishra (2003), Rastogi and Mehrotra (1990, 1991, 1993), Shah and Gopal (1986), Sharma (2003), Singh and Pandey (1998).

Ailanthus triphysa (Dennst.) Alston

Syn.: *A. malabarica* DC.; *A. kurzii* Pram.; *Adenanthera triphysa* Dennst.

Family: Simaroubaceae

Sans. Names	:	*Sans.*: Aralu, Atisarahita, Mahanimbu, Peru.

Distribution	:	Often planted in S. India, Konkan, Deccan, W. Ghat upto 1,000m.
Part/s Used	:	Resin.
Utilization	:	Powdered resin bark taken in small doses with milk for bronchitis.
Active Constituents	:	Trunk exudate possesses malabaricol, epoxy-malabaricol and malabaricanediol.
References	:	Ambasta (1986), Kirtikar and Basu (1984), Rastogi and Mehrotra (1990).

Ajuga bracteosa Wall. *ex* Benth.

Family: Lamiaceae

Distribution	:	Himalaya: 700–1,000m.
Part/s Used	:	Leaves.
Utilization	:	Leaf decoction or its infusion prescribed with honey and ginger juice for respiratory congestion and bronchitis.
Active Constituents	:	Leaves yield glycosides, tannins, ceryl alcohol, β-sitosterol, palmitic acid, triacontanyldocosanoate and tetracosanoic acid.
References	:	Bhattacharjee (1998), Chhetri (1994), Rastogi and Mehrotra (1995), Sharma (2003), Watt (1972).

Alangium salvifolium (L.f) Wangerin

Syn.: *A. lamarckii* Thwaites.; *Grewia salvifolia* L.f.; *A. salvifolium* ssp. *decapetalum* (Lam.) Wangerin; *A. decapetalim* Lamk.; *A. tomentosum* Lamk.

Family: Alangiaceae

Sans. and H. Names	:	*Sans.*: Ankola;
		H.: Akola.
Distribution	:	Throughout the drier parts of India; grows vigorously in the forests of S. India.
Part/s Used	:	Bark.
Utilization	:	10-30 ml decoction of bark given, once a day for 10-15 days, to cure respiratory ailments.
Active Constituents	:	Stem-bark posseses emetine, cephaeline, psychotrine, lamarckinine, alancine and chotrine.
References	:	Chopra *et al.* (1956), Raju (2000), Rastogi and Mehrotra (1990, 1991, 1993), Sharma (1999).

Albizia amara (Roxb.) Boivin.

Syn.: *Mimosa amara* Roxb.; *A. amara* and *Wighti* (Grah.);
M. *pedchella* Roxb.

Family: Fabaceae

Sans. Name	:	*Sans.*: Krishnasirisha.
Distribution	:	Throughout the Himalaya in dry forest. Deccan.
Part/s Used	:	Leaves, Flowers and Seeds.
Utilization	:	Consumption of flower and seed cures asthma. Powdered leaves taken with ghee to cure cough.
Active Constituents	:	Main ingredients of seeds are echinocystic acid, β-sitosterol, an alkaloidal fraction (XI)–budmunchia-mines A, B, C, D, E, F, G, H and I.
References	:	Rastogi and Mehrotra (1990, 1995), Upadhyay and Chauhan (2000), Watt (1972).

Albizia lebbeck (L.) Benth.

Syn.: *Acacia lebbek* (L.) Willd.; *Mimosa lebbeck* L.; *A. speciosa* Willd;
A. *sirissa* Ham.; *Albizia latifolia* Boivin.

Family: Fabaceae

Eng., Sans. and H. Names	:	*Eng.*: Parrot tree, Sizzling tree;
		Sans.: Bhandi, Bhandika, Kalinga;
		H.: Kalsis, Shirish, Siras, Sirsa.
Distribution	:	Throughout India.
Part/s Used	:	Flowers and Bark.
Utilization	:	Decoction of flowers and bark good for asthma.
Active Constituents	:	Principle constituents are lebbeckanins D, F, G and H, benzyl acetate, benzyl benzoate and crocetin (flowers), cardenolide glycosides and antraquinones (stem-bark).
Biological Activity	:	Anti-asthmatic activity found positive.
References	:	Duke *et al.* (2002), Kirtikar and Basu (1984), Rastogi and Mehrotra (1990, 1991), Sharma (2003), Sinha and Sinha (2001), Watt (1972).

Alhagi camelorum Fisch.

Syn.: *A. pseudoalhagi* Desv.; *Hedysarum pseudoalhagi* Bieb.;
A. *maurorum* sensu Baker

Family: Fabaceae

Eng., Sans. and H. Names	:	*Eng.*: Arabian manna plant, Camel thorn;

	Sans.: Adhikantaka, Ananta, Duramula;
	H.: Javansa, Javasa, Yavasa.
Distribution	: Gujarat. N. W. India.
Part/s Used	: Whole Plant and Flowers.
Utilization	: Plant smoked along with black datura, tobacco and ajwain seeds as a remedy for asthma in Konkan region of India. Flowers considered good in asthma.
Active Constituents	: (+) catechin, gallocatechin, (-) epigallocatechin, leucodelphinidin isolated from aerial parts.
References	: Bennet (1987), Kirtikar and Basu (1984), Rastogi and Mehrotra (1993).

Allium cepa Linn.
[Pl. 1 B]
Family: Liliaceae

Eng., Sans. and H. Names :	*Eng.*: Onion;
	Sans.: Dirghapatra, Mahakanda, Nripapriya;
	H.: Piyaz.
Distribution	: Grown throughout the various parts of India.
Part/s Used	: Bulb.
Utilization	: 1tsp extract of onion given, twice a day, for relieving cough in infants.
Active Constituents	: Principle componenets are polyphenols, protocatechuic acid, caffeic, ferulic, acid, quercetin, carbohydrates, sterols, sterol glycosides and propenyl sulfenic acid.
Biological Activity	: Quercetin exhibits anti-asthmatic activity.
References	: Asolkar *et al.* (1992), Kirtikar and Basu (1984), Prasad and Abraham (1984), Rastogi and Mehrotra (1990, 1993), Sharma (2003), Sharma and Singh (2001), Sinha and Sinha (2001).

Allium sativum Linn.
Family: Liliaceae

Eng., Sans. and H. Names :	*Eng.*: Garlic, Poor man's treacle;
	Sans.: Arishtha, Bhutabhna, Katukanda, Lashuna;
	H.: Lahsun, Lasan.
Distribution	: Widely cultivated in India.
Part/s Used	: Bulb.

Utilization	:	Extract and its preparations good in pulmonary phthisis, gangrene of lung, whooping cough and laryngeal tuberculosis.
Active Constituents	:	Important ingredients isolated form bulb are allin, antibiotic principles like allistatin I, allistatin II, garlicin, diallyl trisulphide,diallyl sulphide, allylmethyl-disulphide, diallyl disulphide and protoeruoside B.
Biological Activity	:	Allin inhibited multiplication of *Tuberculus bacilli* in presence of allinase.
References	:	Ambasta (1986), Asolkar *et al.* (1992), Bhalla *et al.* (1982), Bhattacharjee (1998), Gill *et al.* (1993), Kirtikar and Basu (1984), Pandey (2003), Rastogi and Mehrotra (1993, 1995).

Aloe barbadensis Mill.

Syn.: *A. vera* Tourn. *ex* Linn.; *A. vulgaris* (Bauhin) Lamk.; *A. indica* Royle; *A. littoralis* Koening

Family: Liliaceae

Eng., Sans. and H. Names	:	*Eng.*: Barbados aloe, Indian aloe;
		Sans.: Ghrita- Kumari;
		H.: Ghee- Kunvar.
Distribution	:	Grown throughout Indian gardens.
Part/s Used	:	Leaves.
Utilization	:	Powdered inner leaf taken for tuerculosis.
Active Constituents	:	Main constituents of leaves are aloesin, aloerone, β-sitosterol, baraloin, mixture of polysaccharides, malic, citric and tartaric acids.
Biological Activity	:	Leaf extract found to inhibit the growth of *Mycobacterium tuberculosis*; activity attributed to barbaloin.
References	:	Ambasta (1986), Anonymous (1999), Duke *et al.* (2002), Rastogi and Mehrotra (1993, 1995, 1991), Watt (1972).

Alpinia calcarata Rosc.

Family: Zingiberaceae

Distribution	:	Cultivated in Indian gardens.
Part/s Used	:	Rhizome.
Utilization	:	10g powdered rhizome given for bronchial catarrh, twice a day.
References	:	Tripathi and Goel (2001), Watt (1972).

Alpinia galanga (L.) Willd.

Syn.: *Amomum galanga* Lour.; *A. galanga* Sweet.

Family: Zingiberaceae

Eng. and Sans. Names	:	*Eng.*: Greater Galangal;
		Sans.: Aruna, Kulanjan.
Distribution	:	Tropical areas of S. and E. India; Cultivated.
Part/s Used	:	Rhizome.
Utilization	:	Decoction of powdered rhizome given for bronchial catarrh.
Active Constituents	:	Principle ingredients of rhizome are oil, tannins, kaempferide, galangin, alpinin, di-(p-hyroxy-cis-syryl) methane and p-hydroxycinnaldehyde.
Remarks	:	Steam volatile oil of rhizome stimulates bronchial glands when exposed to its vapours.
References	:	Bhattacharjee (1998), Jain (1968), Rastogi and Mehrotra (1995), Sastry (1961), Tripathi and Goel (2001), Thakur *et al.* (1989), Watt (1972).

Alpinia sericeum Roxb.

Family: Zingiberaceae

Part/s Used	:	Seeds.
Utilization	:	10g powdered seeds taken twice daily for cough and cold.
References	:	Tripathi and Goel (2001).

Alstonia scholaris R. Br.

Family: Apocynaceae

Eng., Sans. and H. Names	:	*Eng.*: Dita bark, Devils tree;
		Sans.: Bahuparna, Chatraparna, Saptaparna;
		H.: Chatiam, Satni, Satwim.
Distribution	:	Widely cultivated throughout tropical moist deciduous and tropical forest of India.
Part/s Used	:	Bark, Leaves and Latex.
Utilization	:	Powdered dried leaves smoked to cure chronic cough and asthma. Decoction of bark alone or with leaves used for asthma and bronchitis. Besides, aqueous extract of bark latex used against tuberculosis in Assam.

Active Constituents	:	These are echitamine, echidamidine, echitenine, ditamine, lupeol acetate, β-amyrin acetate, (-) macrocarpamine (bark), diacetyl picraline, lagunamine, angustiloine acid, losanine, alstonamine, alschomine and rhazimanine (leaves).
Biological Activity	:	Aqueous extract of leaves exhibit significant antimicrobial activity against *Mycobacterium tuberculosis*.
References	:	Ambasta (1986), Asolkar *et al.* (1992), Greenish (1999), Kirtikar and Basu (1984), Kothari and Londhe (1999), Lalramnghinglova (1996), Rastogi and Mehrotra (1991, 1995), Satapathy and Brahmam (1996), Sinha and Sinha (2001), Thakur *et al.* (1989), Watt (1972).

Althaea officinalis Linn.

Family: Malvaceae

Eng. and H. Names	:	*Eng.*: Bread and cheese, Guimauve, Marsh Mallow;
		H.: Gulkhairo, Khaira- Kajlwr.
Distribution	:	Naturalized in India: Punjab to Kashmir.
Part/s Used	:	Roots. Flowers.
Utilization	:	Decoction of root boiled in wine or milk and with a pinch of sugar considered a popular remedy for cough, bronchitis, whooping cough, asthma and pleurisy. Infusion of flower given to cure bronchial catarrh.
Active Constituents	:	These are palmitic, myristic, stearic, oleic, linoleic, lauric acid, β-sitosterol, stigasterol, tiliroside, naringenin–4′– β–D–glucoside, dihydrokaempferol–4–glucoside, caffeic, p -hydroxyenzoic, ferulic, coumaric, vanillic and chlorogenic acids, quercetin-3-glucoside, astragalin and populin (flowers), sulphate-isoscutellarein-4′-methyl ether-8-(2″SO_3K) glucoside (I), mucilage, starch, asparagin and a substance allied to lecithin.
References	:	Bhattacharjee (1998), Greenish (1999), Kirtikar and Basu (1984), Rastogi and Mehrotra (1993, 1995), Watt (1972).

Alysicarpus bupleurifolius (L.) DC.

Family: Fabaceae

Distribution	:	Throughout plains of India, ascending to 1200m in hills.
Part/s Used	:	Roots.
Utilization	:	10-30 ml root extract given, twice a day for 7-9 days, for asthma.
References	:	Rastogi andMehrotra (1995), Sharma (1999).

Alysicarpus rugosus (Willd.) DC.

Family: Fabaceae

Part/s Used	:	Leaves. Roots.
Utilization	:	Hot decoction of leaves and root given for cough and cold.
Reference	:	Gill and Nyawuane (1994).

Alysicarpus vaginalis (L.) DC.

Syn.: *A. vaginalis* var. *nummularifolius* (Willd.) Miq.

Family: Fabaceae

Eng. Name	:	*Eng.*: Alyce clover.
Distribution	:	Throughout India, ascending to 1,350m in the N.W. states.
Part/s Used	:	Stem and Leaves.
Utilization	:	Decoction of stem and leaves useful for cough, bronchitis and pneumonia.
References	:	Bennet (1987), Chopra *et al.* (1956), Singh and Pandey (1998).

Amaranthus spinosus L.

Family: Amaranthaceae

Eng., Sans. and H. Names	:	*Eng.*: Prickly amaranth, Spiny pigweed;
		Sans: Alpamarisha, Bahuvirya, Bhandira, Granthila;
		H.: Cholai, Kantenatia.
Distribution	:	Throughout India.
Part/s Used	:	Leaves. Seeds.
Utilization	:	Used as cataplasm in cough, asthma and pneumonia.
Active Constituents	:	Cellulose, fats, oil, inulin, lignin, saponin, tannins, hentriacontane and α-spinasterol isolated from leaves.
References	:	Gill *et al.* (1993, 1997), Kirtikar and Basu (1984), Rastogi and Mehrotra (1991, 1993), Watt (1972).

Amomum sericeum Roxb.

Family: Zingiberaceae

Part/s Used	:	Seeds.
Utilization	:	About 10g powdered seeds taken twice daily for curing asthma and pulmonary affections by the Adi tribe.
Reference	:	Tripathi and Goel (2001).

Amorphophallus campanulatus (Roxb.) Bl.

Syn.: *Arum companulatum* Roxb.; *A. paeoniifolius* (Dennst.) Nicolson; *Dracontium paeoniifolius* Dennst.

Family: Araceae

Eng., Sans. and H. Names	:	*Eng.*: Elephant- root yam;
		Sans.: Kanda, Kandala, Sukandi, Vatari;
		H.: Kanda.
Distribution	:	Cultivated throughout the plains of India.
Part/s Used	:	Tuber.
Utilization	:	Roasted tuber prescribed in hot ashes with oil and salt for asthma and bronchitis.
Active Constituents	:	Tubers yield triacontane, lupeol, etulinic acid, stigasterol, β-sitosterol and its palmitate, glucose, galactose, rhamnose and xylose.
References	:	Asolkar *et al.* (1992), Bennet (1987), Caius (2003), Kitikar and Basu (1984), Rastogi and Mehrotra (1991), Watt (1972).

Anacardium occcidentale Linn.

Family: Anacardiaceae

Eng., Sans. and H. Names	:	*Eng.*: Cashew nut;
		Sans.: Agnikrita, Arushkara, Parvali, Sophara;
		H.: Kaju.
Distribution	:	Naturalized and cultivated especially near the coast in India.
Part/s Used	:	Leaves and Stem-bark.
Utilization	:	Decoction of leaves and stem-bark used against asthma.
Active Constituents	:	These are β-sitosterol, stigmasterol, campesterol and cholesterol (stem-bark), myricetin, agathisflavone, robustaflavone, amentoflavone, quercetin, kaempferol, apigenin, quercetin-3-*O*-rhamnoside, quercetin-3-*O*-glucoside, p-hydroxybenzoic, protocatechuic, gentisic, gallic acid along with glucosides, rhamnosides, arabinosides, xylosides of kaempferol and quercitol (leaves).
References	:	Gill *et al.* (1993), Kirtikar and Basu (1984), Rastogi and Mehrotra (1991, 1995).

Andrographis paniculata (Burm f.) Wall. *ex* Nees
Syn.: *Justicia paniculata* Roxb.
Family: Acanthaceae

Eng., Sans. and H. Names	:	*Eng.*: Creat;
		Sans.: Bhumimba, Kirata;
		H.: Kirayat, Mahatita.
Distribution	:	Throughout India, sometimes cultivated.
Part/s Used	:	Leaves, Seeds and Whole Plant.
Utilization	:	Decoction of leaf and seed taken with a little honey, thrice a day, for cough and cold. Powdered plant prescribed with goat milk for 40 days to cure tuberculosis.
Active Constituents	:	Leaves rich in β–Sitosterol–glucoside, an unsaturated ketone, a diterpene glucoside -deoxyandrographolide–19β–D–glucoside and andrographolide.
References	:	Ambasta (1986), Chopra *et al.* (1956), Rastogi and Mehrotra ((1990,1993), Seetharam *et al.* (1999), Sharma (2003), Singh *et al.* (1997), Sinha and Sinha (2001), Thakur *et al.* (1989), Watt (1972).

Aneilema scapiflorum Weight
Family: Commelinaceae

H. Name	:	*H.*: Siyahmusli.
Distribution	:	Spread from upper Gangetic plains to Bhutan.
Part/s Used	:	Root-bark.
Utilization	:	Dried root-bark employed in asthma.
Reference	:	Caius (2003).

Angelica archangelica Linn. var. *limalaica* (C.B.Clarke) Krishna and Badhuar
Syn.: *Archangelica officinalis* Hoffm. var. *limalaica* C. B. Clarke
Family: Apiaceae

Part/s Used	:	Roots.
Utilization	:	Decoction of root good for bronchial colds.
Active Constituents	:	Roots yield furocoumarins- angelican, prangolarin, archangelin; a phenolic compound–angelicain, flavonoid–archangeliain and dihydrofuranocoumarin glycosides VI, VII.

References	:	Ambasta (1986), Rastogi and Mehrotra (1990, 1991, 1993).

Angelica glauca Edgew.

Family: Apiaceae

Eng. and Sans. Names	:	*Eng.*: Himalayan angelica;
		Sans.: Choraka.
Distribution	:	Kashmir to Shimla: 2,700 to 3,350m.
Part/s Used	:	Roots.
Utilization	:	Used for the treatment of bronchitis.
Active Constituents	:	Various constituents isolated from roots are essential oil (1.3 per cent), resin, valerianic acid, angelicin, bergapten, xanthotoxin, umelliferene, lactone angeolide, furocoumarin–2″–O–acetyloxypeucidanin hydrate, archangelin and oxypeucedanin.
References	:	Bhattacharjee (1998), Rastogi and Mehrotra (1993, 1995), Sharma (2003), Singh (1999), Watt (1972).

Anisochilus carnosus Wall.

Family: Lamiaceae

H. Names	:	*H.*: Kurkka, Panjirikapat, Pathukurkka.
Distribution	:	W. Himalaya. Bengal. C. India. Deccan.
Part/s Used	:	Whole Plant. Leaves.
Utilization	:	Plant used as an expectorant for curing cough. Juice of fresh leaves mixed with sugar given for cough and cold.
Active Constituents	:	Diosmetin and diosmin isolated from leaves.
References	:	Ambasta (1986), Chopra *et al.* (1956), Kirtikar and Basu (1984), Rastogi and Mehrotra (1991).

Anogeissus latifolia Wall. *ex.* Guill. & Perr.

Syn.: *Conocarpus latifolia* Roxb.; *A. latifolia* var. *glabra* Cl.; *A. l* var. *villosa* Cl.; *A.l* var. *tomentosa* Haines

Family: Combretaceae

Eng. and H. Names	:	*Eng.*: Button tree, Dindigo tree, Ghatti tree;
		H.: Bakla, Dhaura.
Distribution	:	Sub–Himalayan tract; ascending to 1,000m in C. and S. India.
Part/s Used	:	Stem-bark.

Utilization	:	1-3 tsp juice of stem-bark (thrice a day for a week) found to be effective by Konda Reddis and Koyas to check cough and asthma.
Active Constituents	:	Stem-bark yields gallic acid, 3,3'- di–O-methyl ellagic acid, its 4-β-D-glucoside, 3, 4, 3'- tri–O–methyl elagic acid, quercetin and myricetin.
References	:	Bennet (1987), Chopra *et al.* (1956), Kumar and Pullaiah (1998), Rao and Henry (1996), Rastogi and Mehrotra (1993), Watt (1972).

Apium graveolens Linn.

Family: Apiaceae

Eng., Sans. and H. Names	:	*Eng.*: Celery, Cultivated Celery, Willd Celery;
		Sans.: Ajamoda, Brahmamusha, Mayura;
		H.: Ajmud, Ajmoda, Bari.
Distribution	:	N.W. Himalaya, largely cultivated.
Part/s Used	:	Seeds.
Utilization	:	Seeds considered good against bronchitis and asthma.
Active Constituents	:	Active ingredients of seeds are celerin, p-hydroxycinnamic acid, celereoside, isoquercitrin, β-carotene, vit. A, palmitate, vit. K, α-tocopherol, seselin, ergapten, rutaretin, celereoin, apiumoside, vallein and nodakenin.
References	:	Das and Agarwal (1991), Thakur *et al.* (1989), Kirtikar and Basu (1984), Rastogi and Mehrotra (1991, 1995).

Aplectrum hyemale (Muhl.) Torr.

Family: Orchidaceae

Distribution	:	Temperate region.
Part/s Used	:	Bulb.
Utilization	:	Bulb used against bronchial problems.
Reference	:	Bhattacharjee (1998).

Apocynum cannabinum Linn.

Family: Apocynaceae

Eng. Name	:	*Eng.*: Black Indian Hemp.
Part/s Used	:	Whole Plant.
Utilization	:	Aqueous extract of plant given for asthma.
Reference	:	Bhattacharjee (1998).

Arctium minus (Hill.) Bernh.

Family: Asteraceae

Part/s Used	:	Leaves. Roots.
Utilization	:	Decoction of leaves prescribed for lungs diseases, cough and asthma.
Reference	:	Bhattacharjee (1998).

Areca catechu Linn.

Family: Arecaceae

Eng., Sans. and H. Names	:	*Eng.*: Areca palm, Betel nut tree, Cashoo nut tree;
		Sans.: Akota, Chikkana, Ghenta, Gopadala, Khapura;
		H.: Supari, Supyari, Suppari.
Distribution	:	Cultivated exclusively with in moist tropical tracts of coastal India.
Part/s Used	:	Leaves.
Utilization	:	Macerated leaves in alcohol given for bronchitis and cough.
References	:	Caius (2003), Greenish (1999).

Argemone mexicana Linn.

Family: Papaveraceae

Eng., Sans. and H. Names	:	*Eng.*: Mexican poppy, Prickly poppy, Yellow Mexican poppy;
		Sans.: Hemavati, Kanchini;
		H.: Bharband, Branhmi, Brahmadundi.
Distribution	:	Throughout India upto 1,700m.
Part/s Used	:	Root and Seeds.
Utilization	:	Decoction of roots and seeds prescribed for cough, pulmonary diseases and asthma. Flower extract considered good for whooping cough. Decoction of stem and leaves given, thrice a day, for chronic cough and asthma.
Active Constituents	:	Roots possess norsanguinarine, ererine, protopine, β–sitosterol, bererine, protopine, argemexitin and 5, 7-dihydroxychromone-7-neohesperidoside (I).
References	:	Bhalla *et al.* (1982), Bhattacharjee (1998), Das (1997), Das and Agarwal (1991), Kirtikar and Basu (1984),

Kumar and Jain (1998), Rastogi and Mehrotra (1993, 1995), Sen and Batra (1997), Singh and Pandey (1998), Sinha and Sinha (2001).

Arisaema triphyllum (L.) Schott.
Family: Araceae

Part/s Used	:	Corm.
Utilization	:	Used to cure asthma and tuberculosis.
Reference	:	Bhattacharjee (1998).

Aristolochia indica Linn.
Family: Aristolochiaceae

Eng., Sans. and H. Names	:	*Eng.*: The Indian Birthwort;
		Sans.: Arkanula, Ishari mul, Ishvara;
		H.: Isharmul.
Distribution	:	The plant grows wild throughout the low hills and plains of India from Nepal to W. Bengal and S. India.
Part/s Used	:	Leaves.
Utilization	:	Three mature leaves chewed on empty stomach for 21 days to cure asthma. Patient should not take water or food for 2-3 hrs.
Active Constituents	:	Leaves contain aristolochic acid, aristolamide, aristolonitrite, aristolinic acid and methyl aristolochate.
References	:	Ambasta (1986), Hembron (1991), Jha and Varma (1996), Kirtikar and Basu (1984) Thakur *et al.* (1989).

Aristolochia tagala Cham.
Syn.: *A. roxburghiana* Klotzsch.; *A. acuminata* Roxb. (non Lam.) Wight
Family: Aristolochiaceae

Distribution	:	Bengal. Assam. W. Peninsula.
Part/s Used	:	Leaves.
Utilization	:	Paste of leaves prepared in coconut oil applied as a rubifacient on chest for asthma and cough.
Active Constituents	:	Leaves yield aristolochic acids A and C, 7 hyroxyaristolochic acid A, allantoin and 4, 7-dimethyl-6-methoxy-1- tetralone.
References	:	Bennet (1987), Dagar and Dagar (1996), Kirtikar and Basu (1984), Rastogi and Mehrotra (1993, 1995).

Arnebia guttata Bunge

Syn.: *A. tibetana* Kurz.

Family: Boraginaceae

Distribution	:	Kashmir: 2,100-3,600m.
Part/s Used	:	Roots.
Utilization	:	Red dye extracted from roots used for cough.
Active Constituents	:	Deoxyshikonin, acetylshikonin, shikonin, β-hydroxyisovaleryl shikonin and tetracrylshikonin isolated from roots.
References	:	Ambasta (1986), Bennet (1987), Rastogi and Mehrora (1993, 1995).

Artemisia indica Willd.

Syn.: *A. vulgaris* auct. non Linn.

Family: Asteraceae

Distribution	:	Throughout the mountainous districts of India upto 1,700–4,000m.
Part/s Used	:	Leaves.
Utilization	:	10 ml infusion of leaves and flowering tops taken twice daily for asthma.
References	:	Bennet (1987), Chopra *et al.* (1956), Lalramnghinglova and Jha (1997).

Artemisia nilagirica (Cl.) Pamp.

Syn.: *A. vulgaris* auct. (non L.); *A. vulgaris* var. *nilagarica* C.B. Clarke

Family: Asteraceae

Hindi Name:

		H.: Nagdona.
Distribution	:	Throughout India.
Part/s Used	:	Flower heads, Leaves and Shoot.
Utilization	:	Infusion of leaves and flowering tops good against asthma.
References	:	Chopra *et al.* (1980), Jain *et al.* (1997), Jain and Saklani (1992), Rao and Shanpru (1981), Rastogi and Mehrotra (1995).

Artemisia vulgaris Linn.

Syn.: *A. indica* Willd.

Family: Asteraceae

Eng., Sans. and H. Names	:	*Eng.*: Indian wormwood, Mother wort, Mugwort;
		Sans.: Barahipusha, Gutthaka, Nagadamani;
		H.: Gathivana, Mastaru, Nagdona.
Distribution	:	Throughout the hilly districts of India.
Part/s Used	:	Leaves.
Utilization	:	Infusion of leaves given to cure asthma.
References	:	Ambasta (1986), Caius (2003), Kirtikar and Basu (1984).

Asclepias speciosa Torr.

Family: Asclepiadaceae

Part/s Used	:	Roots.
Utilization	:	Tea prepared by adding roots useful for tuberculosis treatment.
Reference	:	Bhattacharjee (1998).

Asclepiodora viridis (Walt) Gray

Family: Asclepiadaceae

Part/s Used	:	Roots.
Utilization	:	Tea prepared from roots given for asthma and shortness of breath.
Reference	:	Bhattacharjee (1998).

Aster amellus Linn.

Syn.: *A. trinervius* Roxb.; *A. aperatoides* Turiz.

Family: Asteraceae

Distribution	:	C. and W. Himalayas: 1,700- 2,350m.
Part/s Used	:	Roots.
Utilization	:	Roots used for cough and pulmonary affections.
References	:	Ambasta (1986), Chopra *et al.* (1956), Rastogi and Mehrotra (1995).

Astragalus psilocentrus Fisch.

Syn.: *A. punjabicus* Sirj.; *A. lasius* Blatter.; *A. subumbellatus* Sensu Baker

Family: Fabaceae

Part/s Used	:	Roots.
Utilization	:	Decoction of roots given to cure cough, cold and chronic bronchitis.
References	:	Bennet (1987), Kapur (1996).

Atropa acuminata Royle *ex* Lindle

Syn.: *A. belladonna* Linn.; *A. belladonna* Sensu Cl.

Family: Solanaceae

Eng. and H. Names	:	*Eng.*: Indian atropa, Indian belladonna;
		H.: Angurshapa, Sagangur.
Distribution	:	Kashmir: 2,000- 3,500m. Also, cultivated in H.P.
Part/s Used	:	Root, Leaf and Aerial Part.
Utilization	:	Powdered roots, decoction of leaves and aerial parts useful against asthma and whooping cough.
Active Constituents	:	These are hyoscyamine (0.72 per cent), hyoscine; 2 flavone glycosides- rutin and kaempferol–3–rhamno-galactoside; kaempferol–7–monoglucoside, quercetin–7–monoglucoside (leaves), atropine, hyoscyamine, hyoscine, cuscohygrine and fatty oil (25 per cent) (roots).
Biological Activity	:	Atropine has stimulatory effect on respiratory system of human body.
References	:	Bennet (1987), Jain (1968), Kaul (1997), Rastogi and Mehrotra (1990), Sharma (2003), Singh (1996), Sinha and Sinha (2001).

Atropa belladona Linn.

Family: Solanaceae

Eng. and H. Names	:	*Eng.*: Black cherry, Deadly night shade;
		H.: Angurshefa, Sagangur.
Distribution	:	W. Himalaya: 2,000-3,650m.
Part/s Used	:	Whole Plant.
Utilization	:	Used for whooping cough, asthma and bronchial troubles.
Active Constituents	:	These are hyocyamine, little atropine and hyoscine.

Biological Activity	:	Atropine is a powerful bronchodilator particularly in atopic asthmatics with increased bronchomotor tone.
Remarks	:	It is used in Unani prepration 'Habul- Khaskhash' for cough and cold.
References	:	Bhattacharjee (1998), Kirtikar and Basu (1984), Rastogi and Mehrotra (1993), Thakur *et al.* (1989).

Atylosia lineata Wt. and Arn.
Family: Fabaceae

Distribution	:	Tirupati (A.P).
Part/s Used	:	Leaves.
Utilization	:	Dried leaves smoked for curing asthma.
References	:	Asolkar *et al.* (1992), Tosh (1996).

Azadirachta indica A. Juss.
Syn.: *Melia azadirachta* Linn.
Family: Meliaceae

Eng., Sans. and H. Names	:	*Eng.*: Indian lilac, Margosa tree, Neem tree;
		Sans.: Arishta, Kaitarya, Nimba, Nimbaka, Niryasa;
		H.: Neem, Nim, Nimb.
Distribution	:	Planted all over India; naturally in Deccan Peninsula.
Part/s Used	:	Bark and Young branches.
Utilization	:	Bark powder taken orally to cure cough. Young branches good for cough and asthma.
Active Constituents	:	These are nimocinol, epoxyazadiradione, gedunin, nimolide, deacetylnimbin, nimbin, deacetylazadirachtinol, azadiractanin, 2, 3'-dehydrosalannol, meliacins- nimocinolide, isonimocinolide (leaves), azadirachtol and nimocin (fruits).
Biological Activity	:	Leaf extract inhibited *Mycobacterium tuberculosis* H37 Rv at a concentration of 1/1000 in 3 weeks *in vitro*.
References	:	Das and Agarwal (1991), Jain (1986), Kirtikar and Basu (1984), Rao (1981), Rastogi and Mehrotra (1993, 1995), Shah (1984), Singh and Pandey (1998), Thakur *et al.* (1989).

Azima tetracantha Lam.

Family: Salvadoraceae

Sans. and H. Names	:	*Sans.*: Kantangur, Kundali, Trikantajata;
		H.: Kantagur- Kamai.
Distribution	:	Konkan. Deccan.
Part/s Used	:	Leaves.
Utilization	:	Juice of leaves taken for relief in cough and asthma.
Active Constituents	:	Leaves possess isoharmentine-3-O-rutinoside, friedelin, glutinol, lupeol, β-sitosterol, azimine, azcarpine and carpine.
References	:	Ambasta (1986), Asolkar *et al.* (1992), Kirtikar and Basu (1984), Rastogi and Mehrotra (1991, 1995).

Bacopa monnieria (Linn.) Pennel

Syn.: *Herpestis monnieria* (L.) H.B. K; *Monniera cuneifolia* Mich.; *Lysimachia monniera* L.; *Gratiola monnieria* L.; *Bacopa monnieri* (Linn.) Wettst.

Family: Scrophulariaceae

Eng., Sans. and H. Names	:	*Eng.*: Thyme heaved Gratiola;
		Sans.: Saraswati;
		H.: Brahmi, Manduka parni.
Distribution	:	Commonly in moist and wet places in different parts of India.
Part/s Used	:	Leaves.
Utilization	:	Leaf juice given to infants in bronchitis.
Active Constituents	:	Leaves rich in brahmine, hespestine, mannitol and saponins.
References	:	Anonymous (1999), Bhattacharjee (2001), Jain (1968), Raju (2000), Rastogi and Mehrotra (1990, 1995), Thakur *et al.* (1989).

Balanites aegyptiaca (Linn.) Delile

Syn.: *B. roxburghii* Planch; *Ximenia aegyptica* Roxb.; *B.ferox* G.Don.

Family: Balanitaceae

Eng., Sans. and H. Names	:	*Eng.*: Desert Date;
		Sans.: Angulidala, Ingudi, Inguda,Ingul;
		H.: Hingam, Hingot.

Distribution	:	Arid zones of India.
Part/s Used	:	Bark, Fruit, Seed and Seed oil.
Utilization	:	2 tsp filtrate of crushed roots taken with ash of *Aristida setacea* for 3 days is a common practice for cure against cough among Koyas. Pulp of fruit medicinally reputed for whooping cough. Seed oil and bark also useful in cough and cold.
Active Constituents	:	These are diosgenin, glucose, xylose and rhamnose (seed kernel), bergapten and (+) marmesin (stem-bark), balanitins-1,-2 and -3, 9 saponins- balanitisins A, B, C, D and E (fruit pulp), isorhamnetin–3–rutinoside and–3 -rhamnogalactoside, a sapogenol–6 -methyldiosgenin and a furastanol saponin- balanitoside (fruit).
References	:	Ambasta (1986), Kirtikar and Basu (1984), Mishra and Dixit (1976), Rao and Henry (1996), Rastogi and Mehrotra (1990, 1991, 1993, 1995), Shekhawat and Anand (1984), Singh and Pandey (1998), Watt (1972).

Baliospermum montanum Muell.

Syn.: *B. axillare* Bl.; *B. polyandrum* Wight.; *Jatropha montana* Willd.

Family: Euphorbiaceae

Sans. and H. Names	:	*Sans.*: Anukula, Dantika, Gunapriya, Jayapala, Makunaka;
		H.: Danti, Hakum, Hakun.
Distribution	:	Bihar, N. Bengal, Chhota Nagpur and Assam.
Part/s Used	:	Leaves.
Utilization	:	Decoction of leaves useful in asthma.
References	:	Bennet (1987), Kirtikar and Basu (1984).

Balanophora polyandra Griff.

Family: Dipterocarpaceae

Part/s Used	:	Aerial Parts.
Utilization	:	Considered useful in asthma and cough.
Active Constituents	:	Aerial parts contain coniferin.
Reference	:	Ambasta (1986).

Bambusa arundinacea Retz.

**Syn.: *B. orientalis* Nees.; *Arundo bambos* Linn.;
Bambos arundinacea Pers.**

Family: Poaceae

Eng., Sans. and H. Names	:	*Eng.:* Spiny Bamboo, Thorny Bamboo;
		Sans.: Dridhakanda, Duraruha, Kantaki;
		H.: Bans, Kantabans, Kattang, Mathans, Tajana, Trinaketu, Vansha, Vanya, Yavaphala.
Distribution	:	Often cultivated in many places in N.W.India and Bengal.
Part/s Used	:	Leaves, Stem and Seeds.
Utilization	:	Infusion of leaves taken for bronchitis. A substance called "Tabashir" or Banslochan (a solid) accumulated in its hollow internodes considered good in tuberculosis. Seeds also said to be useful in tuberculosis, bronchitis and asthma.
References	:	Asolkar *et al.* (1992), Watt (1972).

Bambusa vulgaris Schrad.

Syn.: *B. thouarsii* Kunth.; *B. arundinacea* Aiton.

Family: Poaceae

Distribution	:	Cultivated throughout tropical India.
Part/s Used	:	Leaves. Nodes.
Utilization	:	Decoction of leaves and nodes prescribed in cough and bronchitis. Sap given as remedy for phthisis.
References	:	Caius (2003), Watt (1972).

Barleria cristata Linn.

Syn.: *B. dichotoma* Roxb.; *B. cristata* Willd.

Family: Acanthaceae

Sans. Names	:	*Sans.:* Kuruvaka, Raktamlana, Subhaga.
Distribution	:	Throughout India, found wild on the Sub–tropical Himalaya, Sikkim and Khasia Hills.
Part/s Used	:	Whole Plant, Leaves and Root.
Utilization	:	Decoction of whole plant (3 tsp) given, thrice a day for 6 months, for tuberculosis. Plant also considered useful in bronchitis and asthma. Infusion of roots and leaves useful for cough and also its 5-10ml decoction with

black pepper, twice a day, prescribed for treating cough and flu.

References	:	Ambasta (1986), Jha and Varma (1996), Singh and Kumar (2000), Varma (1997), Watt (1972).

Barringtonia racemosa (Linn.) Roxb.

Syn.: *B. racemosa* Blume.

Family: Euphorbiaceae

Eng., Sans. and H. Names	:	*Eng.*: Indians Oak;
		Sans.: Nipa;
		H.: Ijjul.
Distribution	:	Eastern and W. sea coast of India.
Part/s Used	:	Fruits.
Utilization	:	Fruits considered effective against cough and asthma.
Active Constituents	:	Active principles of fruits are rhamnose and barringtogenol.
References	:	Bhattacharjee (1998), Kirtikar and Basu (1984), Rastogi and Mehrotra (1990), Watt (1972).

Bauhinia acuminata Linn.

Family: Fabaceae

Eng. and Sans. Names	:	*Eng.*: Dwarf white Bauhinia;
		Sans.: Sivamalli.
Distribution	:	Bengal and S. India.
Part/s Used	:	Bark and Leaves.
Utilization	:	Decoction of bark as well as leaves given for asthma.
References	:	Ambasta (1986), Watt (1972).

Benincasa hispida (Thunb.) Cogn.

Syn.: *B. cerifera* Savi; *Cucurbita hispida* Thunb.

Family: Cucurbitaceae

Eng., Sans. and H. Names	:	*Eng.*: Ash gourd;
		Sans.: Kooshmanda;
		H.: Petha.
Distribution	:	Cultivated more or less throughout the plains of India and on the hills upto1,350m.
Part/s Used	:	Fruits.

Utilization	:	Decoction of fruits used in respiratory affections.
Active Constituents	:	Fruits rich in lupeol, β-sitosterol and their acetate, in-triacontanol, mannitol, arginine, asparatic acid, glutamic acid, asparagines, glutamine, praline, hydroxyproline, isoleucine, cysteine, l-leucine, glucose and rhamnose.
References	:	Ambasta (1986), Asolkar *et al.* (1992), Bennet (1987), Chopra *et al.* (1956), Rastogi and Mehrotra (1991, 1995).

Bergenia ciliata (Haw.) Sternb.

Syn.: *B. ligulata* Engl.; *Saxifraga ligulata* Wal; *Megarea ciliata* Haw.

Family: Saxifragaceae

Sans. and H. Names	:	*Sans.*: Ashmagha, Giribhita, Silabheda;
		H.: Pakhanabheda, Patharcua.
Distribution	:	Throughout the temperate Himalaya from Bhutan to Kashmir: 2,000-3,000m. Khasia Hills upto 1,200 m.
Part/s Used	:	Roots.
Utilization	:	Herbal tea prepared from roots (50ml) along with sugar candy or with honey (twice a day for a week) given for asthma, cough, cold and flu.
Active Constituents	:	(-) afzelechin, bergenin, its C- glucoside, β-sitosterol and (+) catechin-3-gallate have been isolated from roots.
References	:	Ambasta (1986), Anonymous (1999), Bennet (1987), Kirtikar and Basu (1984), Kumar (2002), Kumar and Rao (2001), Kumar and Singh (2000), Lal *et al.* (1996), Megoneitso and Rao (1983), Rastogi and Mehrotra (1990, 1991), Sharma (2003), Singh (1996).

Betonica officinalis Linn.

Family: Lamiaceae

Part/s Used	:	Leaves.
Utilization	:	Powdered leaves prescribed for chronic bronchitis.
Reference	:	Bhattacharjee (1998).

Betula utilis D. Don

Family: Betulaceae

Eng., Sans. and H. Names	:	*Eng.*: Birch tree;
		Sans.: Bahupata, Bahutvaka, Bhurja, Chitrat;
		H.: Bhejpattra, Bhujpatar, Bhujpattra.

Distribution	:	Temperate Himalaya: Kashmir (2,350-4,000m) to Sikkim (3,000-4, 650m).
Part/s Used	:	Bark.
Utilization	:	Infusion of bark useful for bronchitis.
Active Constituents	:	Plant contains betulin, oleanolic acid, acetyle olenolic acid in addition to leucocyananadin in outer bark and polymeric anthocyanidin in inner bark.
References	:	Kirtikar and Basu (1984), Rastogi and Mehrotra (1991), Sharma (2003), Singh (1996).

Biophytum sensitivum (Linn.) Don.
Syn.: *Oxalis sensitivum* Linn.
Family: Oxalidaceae

Sans. and H. Names	:	*Sans.*: Jalapushpa, Lajjaluka, Pitapushap;
		H.: Lajalu, Lakhshana, Zarer.
Distribution	:	Throughout the hotter parts of India.
Part/s Used	:	Leaves.
Utilization	:	Decoction of leaves given for asthma and phthisis.
Active Constituents	:	Leaves contain an insulin like principle.
Biological Activity	:	Aqueous extract inhibits activity of *Mycobacterium tuberculosis*.
References	:	Ambasta (1986), Asolkar *et al.* (1992), Kirtikar and Basu (1984).

Blumea lanceolaria (Roxb.) Druce
Syn.: *B. myriocephala* DC., *B. longifolia* DC.
Family: Asteraceae

Distribution	:	Sikkim, Himalaya and Assam.
Part/s Used	:	Leaves.
Utilization	:	Decoction of leaves taken to cure bronchitis and asthma.
References	:	Asolkar *et al.* (1992), Chopra *et al.* (1956), Lalramnghinglova (1999).

Boerhaavia diffusa Linn.
Syn.: *B. procumbens* Roxb.; *B. erecta* Gaertn.; *B. repens* L.; *B. r.* var. *diffusa* (L.) Hook. *B.r.* var. *procumbens* (Roxb.) Hook.
Family: Nyctaginaceae

Eng., Sans. and H. Names	:	*Eng.*: Hogweed, Pigweed, Spreading Hog weed;

Sans.:   Nilini,   Punarnava,   Raktakanda, Raktapatrika;

H.: Gadapurana, Sant, Sarikinjngi.

Distribution	:	Widely distributed throughout the plains as a weed in drier parts of India.
Part/s Used	:	Roots, Leaves and Whole Plant.
Utilization	:	Decoction of roots as well as plants considered expectorant and anti-asthmatic.
Active Constituents	:	Active principle of roots are hentriacontane, β-sitosterol, ursolic acid, a phytoecdysone–β–ecdysterone and a new C–methyl flavone, 5, 7- dihydroxy -3′, 4′- dimethoxy–6, 8–dimethyl flavone, triacontanol, a purine nucleoside–hypoxanthine–9–arabinofura- noside, 2 rotenoids–boeravinones A and B and boeravinone C.
References	:	Ambasta (1986), Anonymous (1999), Bennet (1987), Bhattacharjee (1998), Jain (1968), Kirtikar and Basu (1984), Kumar and Dwivedi (2000), Mehrotra *et al.* (1995), Raju (2000), Rastogi and Mehrotra (1990, 1991, 1993, 1995), Sharma *et al.* (1979), Singh and Pandey (1998), Thakur *et al.* (1989), Trivedi (2002), Watt (1972).

Bombax ceiba Linn.

Syn.: *B. malabaricum* DC.; *B. heptaphyllm* Roxb.; *Salmalia malabarica* Schott.; *Gossampinus rubra* Ham.

Family: Bombacaceae

Eng., Sans. and H. Names	:	*Eng.*: Cotton tree, Red silk cotton tree;
		Sans.: Apuram, Chirayu, Dirhayu, Kadala;
		H.: Pagum, Ragat semal, Semul, Simal.
Distribution	:	Common upto 1500 m through the hotter parts of India.
Part/s Used	:	Bark. Gum.
Utilization	:	Bark decoction given for cough, and gum to check pulmonary tuberculosis.
Active Constituents	:	Lupeol, β-sitosterol, a naphthaquinone, potassium nitrate, kaempferol and quercetin isolated from stem- bark.
References	:	Kirtikar and Basu (1984), Manandhar (1990), Panthi and Chaudhary (2003), Rastogi and Mehrotra (1990), Sinha and Sinha (2001), Thakur *et al.* (1989), Watt (1972).

Borassus flabellifer Linn.

Family: Arecaceae

Eng., Sans. and H. Names	:	*Eng.*: Brab tree, Char palm, Desert Palm, Fan palm;
		Sans.: Asavadru, Chirayu, Dirghadru, Drumeshvara;
		H.: Tal, Tar, Tarkajhar.
Distribution	:	Throughout India.
Part/s Used	:	Roots, Bark and Fruits.
Utilization	:	Decoction of bark as well as young roots given in diseases of respiratory tract. Palm sugar useful in cough and pulmonary affections. Fruit juice given as tonic for asthmatic patients.
References	:	Asolkar *et al.* (1992), Caius (2003), Singh and Pandey (1998), Kirtikar and Basu (1984).

Boswellia serrata Roxb.

Syn.: *B. thurifera* Roxb. *ex* Flem.; *B. glabra* Roxb.; *B. thurifera* Colebr.; *Libanus thurifera* Colebr.; *B. serrata* var. *glabra* (Roxb.) Bennett

Family: Burseraceae

Eng., Sans. and H. Names	:	*Eng.*: Indian olibanum tree;
		Sans.: Ashvamutri,Gajashana, Karaka, Kunduru;
		H.: Kundur, Salai, Salga, Sali.
Distribution	:	Dry hilly areas throughout the greater part of India except Assam.
Part/s Used	:	Gum. Stem-bark.
Utilization	:	Bark and gum useful in bronchitis, asthma, cough and pulmonary affections.
Active Constituents	:	Main principles are carbohydrates, glucosides and β-sitosterol (bark), arabinose, fructose, idose, and glucuronic acid, digitoxose, rhamnose, xylose, arabinose, galactose and galacturonic acid, a diterpene alcohol- serratol, β- boswellic acid, its acetyl derivative, 11- keto- β- boswellic acid and its acetyl derivative (gum).
Biological Activity	:	Gum expectorant has stimulant effect.
References	:	Ambasta (1986), Bhatt and Sabnis (1987), Kirtikar and Basu (1984), Rastogi and Mehrotra (1990, 1991), Thakur *et al.* (1989), Watt (1972).

Brassica napus Linn.

Syn.: *B. campestris* L. ssp. *napus* (L.) Hook.

Family: Brassicaceae

Eng. and H. Names	:	*Eng.*: Cole seed, Rape;
		H.: Kali sarson, Lahi, Maghi, Toria.
Distribution	:	Cultivated in India from Punjab to Bengal and in parts of Chota Nagpur.
Part/s Used	:	Roots.
Utilization	:	Root extract useful in chronic cough and bronchial catarrh.
References	:	Ambasta (1986), Bennet (1987), Chopra *et al.* (1956).

Brassica oleracea Linn.

Family: Brassicaceae

Eng., Sans. and H. Names	:	*Eng.*: Cabbage, Colewort, Wild cabbage;
		Sans.: Dalamalini, Kebuka, Kemuka;
		H.: Band gobi, Karamkalla, Kobi.
Distribution	:	Cultivated all over India.
Part/s Used	:	Inflorescence.
Utilization	:	Juice useful in chronic cough and bronchial asthma.
Active Constituents	:	3–sophoroside–7–glucosides of kaempferol, quercetin and isorhamnetin.
References	:	Caius (2003), Kirtikar and Basu (1984), Rastogi and Mehrotra (1991).

Brvphytum sensativum DC.

Family: Oxalidaceae

Part/s Used	:	Leaves.
Utilization	:	Decoction of leaves useful in asthma.
Biological Activity	:	Aqueous extract of fresh leaves partially inhibits the growth of *Mycobacterium tuberculosis*.
Reference	:	Kumar (2002).

Butea monosperma (Lamk.) Taub.

Syn.: *B. frondosa* Koen. *ex* Roxb.; *Erythrina monosperma* Lamk.

Family: Fabaceae

Eng., Sans. and H. Names	:	*Eng.*: Bengal Kino, Butea Gum;

Sans.: Bijasneha, Karaka, Kimsuka, Kinshuka;

H.: Dhak, Palas, Tesu.

Distribution	:	All over India upto 1000m.
Part/s Used	:	Flowers, Root, Bark and Gum.
Utilization	:	Poultice of flowers as well as that of roots applied on chest for curing asthma and clearing congested lungs. Gum considered good in cough. Decoction of bark used for cough and cold.
Active Constituents	:	These are free butein (0.37 per cent) and butrin (0.04 per cent), isobutrin, coreoposin, isocoreoposin, sulphurein, monopermoside and isomanospermoside (flowers), glucose, glycine, a glucoside and an aromatic hydroxyl compound (roots).
References	:	Gill and Nyawuane (1994), Greenish (1999), Kirtikar and Basu (1984), Rastogi and Mehrotra (1995), Shah and Singh (1990), Sharma (2003), Sinha and Sinha (2001), Thakur *et al.* (1989).

Byttneria herbacea Roxb.

Family: Stericulaceae

Distribution	:	Orissa, Chennai, Deccan, Bombay and Konkan.
Part/s Used	:	Roots.
Utilization	:	5g powdered roots mixed with *Piper longum,* black peppers and *Trachyspermum ammi* in equal proportions given, twice daily, for curing asthma.
Reference	:	Brahmam and Saxena (1990).

Caesalpinia cristata Linn.

Syn.: *C. banduc* (L.) Roxb; *Guilandia bonduchella* Linn.; *G. bonduc* W. and A.; *C. bonducella* Fleming.

Family: Fabaceae

Eng., Sans. and H. Names	:	*Eng.*: Fever Nut, Indian filbert, Physic Nut;
		Sans.: Kakachika, Prakiriya, Varini;
		H.: Kanja, Kakachin, Kakcha.
Distribution	:	Throughout hotter parts of India upto 760m on the hills. Common in Bengal and S. India.
Part/s Used	:	Seeds and Roots.
Utilization	:	Roots useful against pulmonary tuberculosis. 1-2 tsp of grounded roasted seeds given for asthma.

Active Constituents	:	Seeds contain phytosterinin, bonducin, fatty oil (20-24 per cent), sucrose, α- caesalpin, β-caesalpin, γ- caesalpin and D- caesalpin.
References	:	Das and Agarwal (1991), Gill and Nyawuane (1994), Kirtikar and Basu (1984), Rastogi and Mehrotra (1990, 1993), Shah (1984), Watt (1972).

Caesalpinia pulcherrima Swartz.

Syn.: *Poinciana pulcherrima* Linn.

Family: Fabaceae

Eng., Sans. and H. Names	:	*Eng.*: Barbardoes pride, Peacock flower, Spanish carnation;
		Sans.: Ratnagandhi, Sideshwara, Sidhakya;
		H.: Guletura.
Distribution	:	Commonly occurring in gardens throughout India.
Part/s Used	:	Flowers.
Utilization	:	An infusion of flowers prescribed for bronchitis and asthma.
Active Constituents	:	β-sitosterol, sebacic acid, quercemeritrin, leucodelphinidin, lupeol, sucrose, glucose, fructose, traces of xylose and 3 phenolic compounds isolated from flowers.
References	:	Kirtikar and Basu (1984), Rastogi and Mehrotra (1991), Watt (1972).

Cajanus cajan (Linn.) Millsp.

Syn.: *C. indicus* Spreng.; *Cytisus cajan* Linn.; *C. flavus* DC.; *C. bicolor* and *C. obcordifolia* Singh.

Family: Fabaceae

H. Name	:	*H.*: Arhar.
Distribution	:	Extensively cultivated throughout India even up 2,000m.
Part/s Used	:	Seeds.
Utilization	:	Powdered dried plant in equal proportion with dried flowers of *Nymphaea alba* taken for Asthma. Seeds useful against bronchitis and cough.
Active Constituents	:	Seeds contain globulins, cajanin and concajanin.
References	:	Bennet (1987), Tripathi and Kumar (2003), Watt (1972).

Calotropis gigantea R. Br.

Syn.: *Asclepias gigantea* Willd.

Family: Asclepiadaceae

Sans. and H. Names	:	*Sans.*: Bhanu, Ganarupa, Prabhakara;
		H.: Akand, Lalak, Mudhar.
Distribution	:	Commonly throughout India in warm and dry places.
Part/s Used	:	Leaf, Flower and Root stock.
Utilization	:	Ash of burnt leaf as well as that of flowers with honey given, thrice a day, to cure whooping cough, asthma and cold. Bread made from 'Bajra' kept overnight in hollow of thick rootstock, roasted on fire considered useful against asthma.
Active Constituents	:	These are calotropin and calotropagenin, β- sitosterol (leaves), β-amyrin and stigasterol (flowers).
References	:	Asolkar *et al.* (1992), Banerjee (1996), Bhatt *et al.* (2001), Kirtikar and Basu (1984), Panthi and Chaudhary (2003), Sharma (2003), Verma *et al.* (1995), Watt (1972).

Calotropis procera (Ait.) Ait. f. subsp. *hamiltonii* (Wight.) Ali.

Syn.: *C. hamiltonii* Wall.; *Calotropis procera* auct. non (Ait.) Ait. f.

Family: Asclepiadaceae

Eng., Sans. and H. Names	:	*Eng.*: Madar tree;
		Sans.: Bhanu, Ganarupi, Ravi;
		H.: Ak, Akada, Madar.
Distribution	:	More or less throughout India in warm dry places.
Part/s Used	:	Leaves and Root.
Utilization	:	1 tsp powdered roots or flowers with black peppers given with honey, twice a day, for asthma, cough and cold. Juice of mature yellow leaves put in nostrils to cure cough, cold and flu. Powder resulted by pounding a roasted piece of turmeric kept along with salt in the hole formed by uprooting plant of *Calotropis procera* for 7 days, considered good against asthma.
Active Constituents	:	Main ingredients are cardenolides (leaves), Quercetin-3-rutinoside identified in roots (1.66 per cent), stem (4.82 per cent), leaves (5.01 per cent), flowers (7.63 per cent) and latex (9.74 per cent). Leaves and stem contains β-sitosterol. Flowers contain β-amyrin and stigmasterol.

References	:	Aditya and Ghosh (1988), Bhatt *et al.* (1999), Jain and Srivastava (2001), Kirtikar and Basu (1984), Rastogi and Mehrotra (1993), Sen and Batra (1997), Singh (2002), Singh and Pandey (1998), Sinha and Sinha (2001), Thakur *et al.* (1989), Watt (1972).

Cannabis sativa Linn.
[Pl. 1C]

**Syn.: *C. gigantea* E.; *C. indica* Lamk.; *C. sativa* sensu Hook.f.;
C. sativa var. *indica* Pers.; *C. sativa* ssp. *indica* var. *kafiristanica* (Vavilov)
Small and Cronquist.**

Family: Cannabinaceae

Eng., Sans. and H. Names	:	*Eng.*: Congo tobacco, Hemp, Indian Hemp;
		Sans.: Ajaya, Ananda, Bhanga, Ganjika, Kaunagni;
		H.: Bhaang, Charas.
Distribution	:	Grown throughout India; naturalized in sub–Himalayan tract.
Part/s Used	:	Leaves.
Utilization	:	Indian hemp mixed with tobacco smoked for relief from asthma.
Active Constituents	:	Principle components of leaves are canniprene, cannabispiradienone, cannabispirenone A, cannabispirenone B, α- and β- cannabispiranols, cannithrene 1, cannithrene 2, canniflavonel, cannifavone 2, 5, 4'–dihydroxy 3', 4' dimethoxydihydrostilbene and cannabispirence.
References	:	Anonymous (1999), Bhattacharjee (1998), Sharma and Sood (1997), Sharma (2003), Jain and Tarafdar (1970), Lal and Yadav (1983), Kirtikar and Basu (1984), Rastogi and Mehrotra (1995).

Capparis decidua Edgew.

Syn.: *C. aphylla* Roth; *Sodada decidua* Forsk.

Family: Capparidaceae

Sans. and H. Names	:	*Sans.*: Granthila, Gudhapatra, Kariba, Kataphala;
		H.: Karer, Karel, Kurrel.
Distribution	:	Punjab. C. India. Gujarat. Deccan.
Part/s Used	:	Bark.

Utilization	:	Powdered root-bark with hot water taken for relief from cough and asthma.
Active Constituents	:	Isocodonocarpine, 14–N -acetylisocodonocarpine and 15-N-acetyl- capparisine isolated from root- bark.
References	:	Ambasta (1986), Bennet (1987), Kirtikar and Basu (1984), Rastogi and Mehrotra (1990, 1995), Singh and Pandey (1998).

Capparis divaricata Lamk.

Syn.: *C. stylosa* DC.

Family: Capparidaceae

Part/s Used	:	Roots.
Utilization	:	Powdered roots prescribed with honey for cough, asthma and bronchitis.
References	:	Bennet (1987), Goud and Pullaiah (1996).

Capsicum annuum Linn.

Syn.: *C. frutescens* sensu Cl.

Family: Solanaceae

Eng. and H. Names	:	*Eng.*: Birds Eye Chilli, Paprika, Sweet Pepper;
		H.: Gachmarich, Lalmirich.
Distribution	:	Cultivated as kitchen garden crop throughout India.
Part/s Used	:	Fruits.
Utilization	:	Paste of ripe fruits with pulp of 3 fruits of *Tamarindus indica* and a pinch of common salt given, twice or thrice a day till cure, for common cold.
Active Constituents	:	Capsaicin, capsicin, solanincand, ascorbic acid, homodihydrocapsaicin, lutein, capsorubin, cryptoxanthin, rubixanthin and phytofluene isolated from fruits.
References	:	Asolkar *et al.* (1992), Gogoi and Borthakur (2001), Kirtikar and Basu (1984), Rastogi and Mehrotra (1990, 1993, 1995), Sharma (2003).

Carica papaya Linn.
[Pl. 2A]

Family: Caricaceae

Eng., Sans. and H. Names	:	*Eng.*: Melontree, Papaw, Papaya.;
		Sans.: Chirbuita, Nalikadala;
		H.: Papeeta, Pepiya, Popaiya.

Distribution	:	Cultivated in various parts of India.
Part/s Used	:	Latex.
Utilization	:	A few drops of latex mixed with sugar given for whooping cough and asthma.
Active Constituents	:	Latex contains papain, chymopapain, amylase, lipase and pectase.
References	:	Jain and Sikarwar (1998), Kirtikar and Basu (1984), Sharma (2003), Vieira (1992).

Carissa opaca Stapf *ex* Haines.

Family: Apocynaceae

Part/s Used	:	Roots. Leaves.
Utilization	:	Poultice of paste of its roots along with that of *Bombax ceiba* and rhizome paste of *Acorus calamus* applied on chest and back of the body for curing pneumonia. Leaves also used for cough.
References	:	Maheshwari and Singh (1987), Singh and Prakash (1994).

Carmona microphylla (Lam.) G. Don

Syn.: *Ehretia microphylla* Lam.; *C. retusa* (Vahl) Masamune; *Cordia retusa* Vahl.; *E. buxifolia* Roxb.

Family: Boraginaceae

Part/s Used	:	Leaves.
Utilization	:	Decoction of leaves useful for cough.
Active Constituents	:	α- and β- amyrins and bauerenol isolated.
References	:	Ambasta (1986), Rastogi and Mehrotra (1995).

Casearia esculenta Roxb.

Syn.: *Casearia zeylanica* (Gaertn.) Thw.; *Varica zeylanica* Gaertn.; *C. ovata* (Lamk.) Willd.

Family: Flacourtiaceae

Sans. Names	:	*Sans.*: Bhurigandha, Bhutagandha, Divya, Surabhi
Distribution	:	Konkan and Cochin.
Part/s Used	:	Roots.
Utilization	:	Root considered good for asthma and bronchitis.
Active Constituents	:	Active principles of roots are due to leucopelargonidin, resin and tannic acid.
References	:	Chopra *et al.* (1956), Kirtikar and Basu (1984).

Cassia absus Linn.
Family: Fabaceae

Sans. and H. Names	:	*Sans.*: Chipita, Kulamasha, Kulani, Kulattha;
		H.: Chaksi, Chaksu, Chakut.
Distribution	:	Throughout India.
Part/s Used	:	Leaves.
Utilization	:	Leaves useful for cough, asthma and bronchitis.
Active Constituents	:	Leaves contain quercetin and rutin.
References	:	Kirtikar and Basu (1984), Rastogi and Mehrotra (1991).

Cassia alata Linn.
Family: Fabaceae

Eng., Sans. and H. Names	:	*Eng.*: Ringworm Senna, Winged Senna;
		Sans.: Dadrughna, Dvipagasti;
		H.: Dadmurdan, Datkapat.
Distribution	:	Various parts of India.
Part/s Used	:	Leaves, Flower and Stem-bark.
Utilization	:	Decoction of fresh leaves, flowers and/or Stem-bark good for bronchitis and asthma.
Active Constituents	:	Leaves possess isochrysophanol, physcion-L-glucoside, aloe-emodin, rhein and kaempferol glucoside.
References	:	Bhattacharjee (1998), Gill and Nyawuame (1994), Kumar (2002), Kirtikar and Basu (1984), Rastogi and Mehrotra (1991, 1993), Sharma (2003).

Cassia auriculata Linn.
Syn.: *Senna auriculata* Roxb.
Family: Fabaceae

Eng. and H. Names	:	*Eng.*: Tanner's Cassia, Tanners Senna;
		H.: Taroda, Tarval, Tarvar.
Distribution	:	The Peninsular Rajputana.
Part/s Used	:	Roots and Seeds.
Utilization	:	Roots allays asthma. Seed extract useful for asthma and cough.

Active Constituents	:	Main constituents of roots are 1, 5, 8–trihydroxy–6–methoxy–2–methylanthraquinone–3–O–β–D–galactopyranosyl (1→4)–O–β–D–mannopyranoside.
References	:	Asolkar *et al.* (1992), Kirtikar and Basu (1984), Rastogi and Mehrotra (1995), Singh and Pandey (1980), Watt (1972).

Cassia fistula Linn.
[Pl. 2B]

Syn.:*C. Rhombifolia* Roxb.; *C. fistula* Willd.; *Cathartocarpus fistula* Pers.

Family: Fabaceae

Eng., Sans. and H. Names	:	*Eng.*: Cassia, Golden shower,Indian laburnum;
		Sans.: Aruja, Kritamala, Sampaka, Shefalika;
		H.: Amaltas, Bandarlauri, Girmalah.
Distribution	:	Throughout India upto 1,300m; more common in evergreen forest.
Part/s Used	:	Leaves. Stem-bark. Fruits.
Utilization	:	Extract of stem–bark and leaves used to cure bronchitis and pneumonia. Ash of fruit along with honey (2-3g) useful in whooping cough. Fruit pulp together with half quantity of crystalline sugar taken, once daily, for whooping cough and tuberculosis.
Active Constituents	:	Main ingredients are rhein, mucilage, protein and volatile oil (pulp), anthraquinones derivatives, 7 biflavonoids (I-VII) and 2 triflavonoids (VIII–IX) (leaves), anthroquinone- fistulic acid, sennoside A and B, lupeol, β-sitosterol and hexacosanol (stem-bark).
References	:	Anonymous (1999), Asolkar *et al.* (1992), Bhattacharjee (1998), Das (1997), Jain (1996), Kirtikar and Basu (1984), Rastogi and Mehrotra (1990, 1995), Sharma (2003), Singh and Pandey (1998), Sinha and Sinha (2001), Tarafdar (1987), Watt (1972).

Cassia occidentalis Linn.

Syn.: *Senna occidentalis* Roxb.

Family: Fabaceae

Eng., Sans. and H. Names	:	*Eng.*: Negro coffee, Rubbish Cassia, Stinking Weed;
		Sans.: Arimada, Kanaka, Karkasha;
		H.: Barika sondi, Chakunda, Kasonda.

Distribution	:	Common weed scattered from Himalaya to W. Bengal, S. India.
Part/s Used	:	Leaves and Seeds.
Utilization	:	Extract of leaves and seeds considered a useful remedy for cough by Munda tribe. Powdered roasted seeds also given in cough and whooping cough.
Active Constituents	:	Active principles are sennoside, N-Me-morphidine, anthroquinone glucoside, anthraquinones, polysaccharides-galactomanna, free and glycosides of chrysophanol and physcion, dianthronic heteroside (seeds), dianthronic heteroside, chrysophanol and emodin, free and as glycosides and physcion (leaves).
References	:	Bhatt and Mitaliya (1999), Kirtikar and Basu (1984), Rastogi and Mehrotra (1990, 1991, 1993, 1995), Sharma (2003), Singh and Pandey (1980), Varma (1997).

Cassia sophora Linn.

Syn.: *C. chinensis* Jacq.; *Senna sophera* Roxb.; *S. esculenta* Roxb.; *S. purpurea* Roxb.

Family: Fabaceae

Sans. Names	:	*Sans.*: Kasamarda, Kasaripu.
Distribution	:	Throughout India.
Part/s Used	:	Whole Plant.
Utilization	:	Decoction of plant prescribed for acute bronchitis.
Active Constituents	:	Plant contains emodin and chrysophanic acid.
References	:	Aditya and Ghosh (1988), Chopra *et al.* (1956). Kirtikar and Basu (1984), Singh and Pandey (1998), Watt (1972).

Cassia tora Linn.

Syn.: *C. obtusifolia* Linn.; *Senna toroides* Roxb.; *C. tora* Sensu Baker non L.; *C. tora* var. *obtusifolia* (L.) Haines

Family: Fabaceae

Eng., Sans. and H. Names	:	*Eng.*: Fetid Cassia, Ring worm plant;
		Sans.: Avudham, Chakri, Kusuma, Padmata;
		H.: Chakund, Panevar.
Distribution	:	Throughout India in warmer places ascending to 1,400m.
Part/s Used	:	Leaves, Stem, Fruits and Seeds.

Utilization	:	5-10g of powdered roasted seeds mixed with tea, given, twice daily, for cough. 100ml decoction of seeds and stem in equal proportion, (twice daily for 5-10 days) considered as anti-asthmatic preparation. Decoction of plant taken in acute bronchitis. Leaves and fruits also recommended for asthma and bronchitis.
Active Constituents	:	These are d-mannitol, myricyl alcohol and sitosterol (leaves and stem), anthraquinones, β-sitosterol, emodin, subrofusarin, 1, 8-dihydroxy anthraquinones (roots and seeds).
References	:	Bhattacharjee (1998), Kumar (2002), Jain (1997), Kirtikar and Basu (1984), Manandhar (1990), Mandal and Basu (1996), Sharma (2003).

Castanea sativa Mill.

Syn.: *C. vulgaris* Lam.; *C. vesca* Gaertn.

Family: Fagaceae

Eng. Name	:	*Eng.*: Sweet Chestnut.
Distribution	:	Grown in Punjab, Darjeeling and the Khasia hills.
Part/s Used	:	Leaves.
Utilization	:	Aqueous infusion of leaves useful in paroxysmal cough and irritable condition of respiratory organs.
Active Constituents	:	Narcissin, helichryoside, genkwanin, kaempferol-3-O-[glycosyl (2''-p-coumaroyl)] (6→1)-α-L-rhamnopyranoside, tiliroside, myricetol, quercitol, rutinoside, myricitroside, quercitroside isolated from leaves
Biological Activity	:	Et OH (50 per cent) extract of stem-bark spasmolytic.
References	:	Ambasta (1986), Rastogi and Mehrotra (1991, 1993), Watt (1972).

Castanopsis purpurella (Miq.) Balak.

Syn.: *Castanea purpurella* Miq.; *Castanopsis hystrix* DC.

Family: Fagaceae

Distribution	:	N.E. part of India: 600- 2,700 m.
Part/s Used	:	Leaves.
Utilization	:	Decoction of leaves (25-50ml) useful for cough and bronchitis.
Active Constituents	:	Castanopsinin I, castanopsinins A, C and E isolated from leaves.

References	:	Bennet (1987), Prakash and Mehrotra (1991), Rastogi and Mehrotra (1995).

Catha edulis Forsk.

Family: Celastraceae

Eng. Name	:	*Eng.*: African tea.
Distribution	:	Bombay and Mysore.
Part/s Used	:	Leaves.
Utilization	:	Infusion of leaves prescribed for cough and asthma.
Active Constituents	:	Dihydromyricetin and its 3–O–rhamnoside, cathedulins K17, K19 and K20; d-norisoephedrine, 17 amino acids, kaempferol, quercetin, myricetin and dulcitol isolated from leaves.
References	:	Ambasta (1986), Asolkar *et al.* (1992), Chopra *et al.* (1956), Rastogi and Mehrotra (1993, 1995).

Celastrus paniculatus Willd.

Syn.: *Celastrus alnifolia* Don; *C. dependens* Wall; *C. multiflora* and *mutans* Roxb.

Family: Celastraceae

Eng., Sans. and H. Names	:	*Eng.*: Black oil tree, Intellect tree, Staff tree;
		Sans.: Agnimasha, Durjara, Kanguni, Lavana;
		H.: Malikangatu, Malkakni.
Distribution	:	Sub–Himalayan tract upto 2,000m. Hilly areas of Maharashtra, Gujarat and Tamil Nadu.
Part/s Used	:	Bark and Seed-Oil.
Utilization	:	1 tsp of seed oil given with yolk of one raw egg on empty stomach for 8-30 days to cure cough and asthma. Decoction of bark useful for bronchitis.
References	:	Anonymous (1999), Hembrom (1991), Joshi *et al.* (1980), Kirtikar and Basu (1984), Rastogi and Mehrotra (1995), Thakur *et al.* (1989), Varma (1997), Watt (1972).

Celosia argentea L. var. *cristata* (L.) O. Kuntze

Syn.: *C. cristata* L.; *C argentea* var. *cristata* (L.) Kuntze.

Family: Amaranthaceae

Eng., Sans. and H. Names	:	*Eng.*: Cock's comb, Quail grass;
		Sans.: Madhichaada, Mayurashikha, Vitunna.
		H.: Morashikha, Pilamurghka, Sufaid murgha.

Distribution	:	Throughout India.
Part/s Used	:	Whole Plant.
Utilization	:	Plant ash mixed with honey given to cure asthma and cough.
Active Constituents	:	Various constituents are hordenine, β-siosterol and ancistrocladine, cholesteryl palmitate, 3, 5–dihydroxy–benzaldehyde, 4–hydroxybenzoic acid, 3, 4–dihydroxybenzoic acid, n–butyl–β–D -fructoside and sucrose.
References	:	Bennet (1987), Kirtikar and Basu (1984), Punjani (2002), Rastogi and Mehrotra (1993, 1995).

Centella asiatica (L.) Urban.

Syn.: *Hydrocotyl asiatica* L.

Family: Apiaceae

H. Name	:	H.: Brahmi.
Distribution	:	Throughout India.
Part/s Used	:	Leaves.
Utilization	:	10 ml decoction of leaves mixed with ginger, black pepper and cardamom given, twice daily, for allaying cough, cold and tuberculosis.
Active Constituents	:	Main ingredients are hydrocotyline, asiaticoside, brahamoside, brahminoside, brahmic acid, centric acid, centrillic acid, thankunic acid, ascorbic acid, kaempferol, quercetin, free amino acids, asparatic acid, glutamic acid, glycine, alanine and phenylalanine.
References	:	Kharkongor and Joseph (1981), Kumar (2002) Panthi and Chaudhary (2003), Sharma (2003), Thakur *et al.* (1989).

Cephalenthus occidentalis Linn.

Family: Rubiaceae

Distribution	:	Various parts of India.
Part/s Used	:	Roots.
Utilization	:	Decoction of roots mixed with honey prescribed for pleurisy.
Reference	:	Bhattacharjee (1998).

Cephaelis ipecacuanha A. Rich.

**Syn.: *C. emetica* Pers.; *Callicocca ipecacuanha* Brot.;
Ipecacuanha officinalis Arrundo.**

Family: Rubiaceae.

Eng. Name	:	*Eng.:* Ipecac.
Distribution	:	Cultivated in Darjeeling, Assam and Bengal.
Part/s Used	:	Roots.
Utilization	:	Roots said to be useful for acute bronchitis and whooping cough. Rhizome of the plant constitutes a drug Ipecac, which brings about vomiting for relief in cough.
Active Constituents	:	Roots yield emetin, and cephaline, psychotropine, methyl ether, emetamine, ipecacuanic acid, ipeoside, erythrocephaline, saponin, resin, fats, calcium oxalate and starch.
References	:	Bhattacharjee (1998), Jain (1968), Thakur *et al.* (1989), Watt (1972).

Chasalia chartacea Craib

Syn.: *C. curviflora* Thw.; *Psychotria curviflora* Wall.

Family: Rubiaceae

Distribution	:	E. Tropical Himalaya: 700-2,000m.
Part/s Used	:	Roots.
Utilization	:	Decoction of roots good for cough and pneumonia.
References	:	Ambasta (1986), Chopra *et al.* (1956).

Chelianthes abomarginala Clarke

Family: Sinopteridaceae

Part/s Used	:	Whole Plant.
Utilization	:	Extract of the plant mixed with honey taken after meal by weak people suffering from tuberculosis.
Reference	:	Bhattacharjee (1998).

Chlorophytum borivilianum Sant. and Fern.

Family: Liliaceae

Distribution	:	Sal forest and moist teak forest of M.P, Chhatisgarh, Jharkhand.
Part/s Used	:	Roots.

Utilization	:	Aqueous extract of root taken, twice daily for 7 days, to cure cough.
References	:	Sharma (2003), Chaudhary and Hutke (2002).

Cicer arietinum Linn.
Family: Fabaceae

Eng., Sans. and H. Names	:	*Eng.*: Bengal gram, Chick pea, Common gram;
		Sans.: Chanaka, Jivana, Kanchuki;
		H.: Chana, Chunna.
Distribution	:	Cultivated in most parts of India.
Part/s Used	:	Leaves and Seeds.
Utilization	:	Leaves and seeds useful in bronchitis.
Active Constituents	:	Chief ingredients are butanoic acid lactone (I), kaempferol–3–(malonyl) glucoside, kaempferol–3–(apiosyl–malonyl) glucoside, vanillic and p-hydroxybenzoic acid (leaves), isoaliquiritigenin, isoliquiritigenin- 4'- glucoside, 3,4',7- trihydroxy-flavone, daidzein, pratensein, p-coumaric acid, biochanin–7–glucoside, pyrimidine nucleotides–uracil–5–β–D–fructofuranosyl–1'-monophosphate and uracil-5–β–D–ribofuranosyl–2 '- 3–cyclic monophos-phate and β- sitosterol (seeds).
References	:	Kirtikar and Basu (1984), Rastogi and Mehrotra (1990, 1991, 1993, 1995).

Cimicifuga racemosa (L.) Nutt.
Family: Ranunculaceae

Distribution	:	Temperate Himalaya–Kashmir to Bhutan upto 2,000 m.
Part/s Used	:	Rhizome.
Utilization	:	Used in the form of tincture at early stage of chronic bronchial disease.
Active Constituents	:	Rhizome contains actinea, cimicfugoside, isoferrulic acid, sugar, tannin, phytosterol and racemosin.
References	:	Asolkar *et al.* (1992), Bhattacharjee (2001), Greenish (1999), Watt (1972).

Cinnamomum camphora (L.) Nees & Eberm.

Syn.: *Camphora officinarum* Bauh, *Laurus camphorifera* Kamp.

Family: Lauraceae

Eng. and Sans. Names	:	*Eng.*: Camphor, Gum camphor;
		Sans.:Karpura.
Distribution	:	Cultivated in Nilgiri Hills of India.
Part/s Used	:	Twigs and Leaves.
Utilization	:	Oil extracted from leaves and twigs useful in bronchitis.
Active Constituents	:	Leaves possess sesquiterpenoids–9-oxoneroldiol, cis- and trans- 3,7,11-trimethyldodeca-1,7,10- trien-3-ol-9-one and 9-oxofarnesol, lignans- kukunokinin, cinnamonol and kukunokinol alng with dimethyl matairesinol, hinokinin and dimethyl secoisolariciresinol.
References	:	Bhattacharjee (1998), Kirtikar and Basu (1984), Rastogi and Mehrotra (1991), Sharma (2003), Watt (1972).

Cinnamomum glanduliferum Meissn.

Syn.: *Laurus glandulifera* Wall.

Family: Lauraceae

Distribution	:	C. Himalaya. Khasia hills.
Part/s Used	:	Leaves and Bark.
Utilization	:	Decoction of young leaves and inner bark is useful against cough and cold.
References	:	Chhetri (1994), Kirtikar and Basu (1984), Watt (1972).

Cinnamomum ineris Reiwn.

Syn.: *Laurus nitida* Roxb.

Family: Lauraceae

H. Names	:	*H.*: Darchini, Janglidarchini.
Distribution	:	Evergreen forest of W. Ghat from Mysore, Coorg to Anamalais.
Part/s Used	:	Seeds.
Utilization	:	Bruised seeds mixed with honey said to be good for cough.
References	:	Das and Agarwal (1991), Kirtkar and Basu (1984), Watt (1972).

Cinnamomum tamala Nees & Eberm.

Syn.: *C. albiflorum* Nees; *Laurus cassia* Roxb.

Family: Lauraceae

Eng., Sans. and H. Names	:	*Eng.*: Cassia cinnamon, Cassia Lignea;
		Sans.: Ankusha, Tajpatra, Talispatri,Tejkalam;
		H.: Brahmi, Dalchini, Tamalapatra, Tejpat.
Distribution	:	Sub-tropical part of Himalaya: 1,000-2,000m but common at 1,000-1,200m.
Part/s Used	:	Bark and Leaves.
Utilization	:	Powdered leaves and bark mixed with tea for cough, cold and bronchitis.
Active Constituents	:	Essential oil contains inalool, cinnamaldehyde, kaempferol and quercetin.
References	:	Kharkongor and Joseph (1981), Kirtikar and Basu (1984), Kumar (2002), Rana *et al.* (1996), Sinha and Sinha (2001), Thakur *et al.* (1989), Watt (1972).

Cissampelos pareira Linn.

Syn.: *C. hernandifolia* Wall; *C. hirsuta* Buch–Ham *ex* DC.; *C. nepalensis* Rhodes; *Menispermum obriculatum* L.

Family: Menispermaceae

Eng. and Sans. Names	:	*Eng.*: Pareira Brava;
		Sans.: Patha.
Distribution	:	Tropical and Sub–tropical regions.
Part/s Used	:	Roots.
Utilization	:	Powdered roots or its extract useful for asthma, bronchitis and pneumonia.
Active Constituents	:	Roots yield hayatin, L-curine and d-isochondro-dendrine, pelosine or bebeerines (0.5 per cent).
References	:	Bennet (1987), Chopra *et al.* (1956), Das (1997), Jain (1996), Rastogi and Mehrotra (1990), Watt (1972).

Cissus quadrangularis L.

Syn.: *Vitis quadrangularis* Wall.; *Cissus edulis* Dalz.

Family: Vitaceae

Eng., Sans. and H. Names	:	*Eng.*: Adamant creeper, Bone Setter, Edible- Stemmed
		Sans.: Amara, Astisamhara, Kandalata;
		H.: Hadjora, Kandavel, Nallar.

Distribution	:	Throughout the hotter parts of India.
Part/s Used	:	Stem. Leaves.
Utilization	:	Tender stems and leaves eaten as vegetable for bronchitis by Jatapus and Savarus. A paste of stem also prescribed in asthma.
Active Constituents	:	Stem yields 3-ketosteroid.
References	:	Bhandary and Chandrashekhar (2002), Chopra *et al.* (1956, 1968), Kirtikar and Basu (1984), Kumar (2002), Rao and Henry (1996),

Citrullus colocynthis Schrad.

Syn.: *Cucumis colocynthis* Linn.

Family: Cucurbitaceae

Eng., Sans. and H. Names	:	*Eng.*: Bitter Apple, Bitter Cucumber;
		Sans.: Atmaraksha, Chitrala, Mahaphala, Ramya;
		H.: Ghorumba, Indrayam, Makal.
Distribution	:	Found throughout India.
Part/s Used	:	Fruits.
Utilization	:	Fruits considered useful for treating asthma and bronchitis.
Active Constituents	:	Fruits rich in cucubitacins B, E, I and cucubitacin E-2 glucoside.
References	:	Arshad *et al.* (1997), Kirtikar and Basu (1984), Rastogi and Mehrotra (1993), Watt (1972).

Citrus maxima Merrill.

Syn.: *Aurantium maximum* Burm.; *C. grandis* (L.) Osbeck.; *Citrus decumana* Murr.

Family: Rutaceae

Eng., Sans. and H. Names	:	*Eng.*: Forbidden fruit, Grape fruit, Paradise apple;
		Sans.: Madhukarkati, Pampalamasi;
		H.: Chakotra, Mahanibu, Sadaphal.
Distribution	:	Grown in Coorg, Mysore, Mumbai, Punjab, Chennai and U.P.
Part/s Used	:	Leaves, Fruits and Gum.
Utilization	:	Leaves are useful in convulsive cough. Gum also considered as a remedy for cough. Fruits considered good in asthma and cough.

Active Constituents	:	Active principles of fruits are 5-geranyloxy-7-methoxycoumarin, soimperatorin, bergapten and 5-geranyloxypsoralen.
References	:	Chopra *et al.* (1956), Kirtikar and Basu (1984), Rastogi and Mehrotra (1995).

Clematis triloba Heyne *ex* Roth.
Family: Ranunculaceae

Sans. and H. Names	:	*Sans.*: Devashreni, Murva.
		H.: Churahar.
Distribution	:	Mumbai, Konkan, Deccan and W. Ghats.
Part/s Used	:	Leaves.
Utilization	:	Leaf decoction useful for asthma.
References	:	Bhattacharjee (1998), Kirtikar and Basu (1984).

Clerodendrum serratum Spreng.
Family: Verbenaceae

Sans. Names	:	*Sans.*: Angaravallari, Barbara, Bhriguja.
Distribution	:	Sub–Himalayan tract, Khasia Hills and S. India.
Part/s Used	:	Roots.
Utilization	:	Roots considered effective in treating asthma and cough.
Active Constituents	:	Roots contain sapogenin, triterpene, starch and resin.
Biological Activity	:	Alcoholic extract from root-bark causes release of histamine from lung tissue.
References	:	Kirtikar and Basu (1984), Rastogi and Mehrotra (1990), Sharma (2003), Watt (1972).

Clerodendrum siphonanthes (R. Br) C.B. Clarke
Syn.: *C. indicum* (Linn.) Kuntze.; *Siphonanthes indica* Linn.
Family: Verbenaceae

Sans. and H. Names	:	*Sans.*: Bhargi, Brahmanayashtika.
		H.: Bharangi, Barangi.
Distribution	:	Chennai, Kumaon and Assam.
Part/s Used	:	Leaves and Roots.
Utilization	:	A glassful juice of leaf along with a little honey diluted in luke warm water taken thrice a day for cough and cold. Roots considered useful in asthma and cough.

Active Constituents	:	Principle components of leaves are diterpene hydroquinone-uncinatone; pectolinarigenin, hispidulin, scutellarein-7-O-glucuronide and hispidulin-7-O-glucuronide.
References	:	Ambasta (1986), Jain (1984), Kirtikar and Basu (1984), Rastogi and Mehrotra (1991, 1995), Singh *et al.* (1997), Tarafdar (1987), Watt (1972).

Clitoria ternatea Linn.

Family: Fabaceae

Eng., Sans. and H. Names	:	*Eng.*: Clitoria;
		Sans.: Aparajita, Bhadra, Girikarnika, Shvela.
		H.: Aparajita, Kajina, Kalina.
Distribution	:	Garden plant found throughout India; wild in Assam and S. India.
Part/s Used	:	Leaves and Roots.
Utilization	:	Pills made by drying 10g root powder with a little hen blood and honey in the sun for 7 days, useful in chronic cough (1 pill, thrice daily, for 15 days). Also, root juice given with cold milk to remove phleg in chronic bronchitis in the Konkan region. Paste of 8–10 leaves crushed with 2-3 black pepper given with water, once daily for 30 days, to cure tuberculosis.
Active Constituents	:	These are 4-kaemferol glycosides- kaemferol-3-glucoside, 3-rutinoside, 3-neohesperidoside and 3-O-rhamnosyl $(1{\rightarrow}2)$- O–rhamnosyl $(1{\rightarrow}2)$–O -[rhamnosyl $(1{\rightarrow}6)$]-glucoside; stigasta-4-en-3, 6-dione, aparijitin (leaves), taraxerol and taraxerone (roots).
References	:	Anonymous (1999), Asolkar *et al.* (1992), Kirtikar and Basu (1984), Thakur *et al.* (1989), Rastogi and Mehrotra (1995).

Cocos nucifera Linn.

Family: Arecaceae

Eng., Sans. and H. Names	:	*Eng.*: Cocoa Nut, Coconut palm, Coconut tree.
		Sans.: Dridhanira, Junga, Magalya, Rasaphala, Shriphala, Trinaraja;
		H.: Nariyal, Narel, Nariel.
Distribution	:	Tropical regions.
Part/s Used	:	Roots.

Utilization	:	Infusion of roots considered useful in bronchitis. Emulsion of oil and kernel prescribed for cough and pulmonary diseases.
References	:	Bhalla *et al.* (1982), Caius (2003).

Coffea arabica Linn.

Family: Rubiaceae

Eng. Name	:	*Eng.*: Coffee.
Distribution	:	Commercially cultivated in India.
Part/s Used	:	Beans.
Utilization	:	Decoction of powdered roasted beans recommended for periodic nervous asthma and whooping cough.
Active Constituents	:	Beans possess caffeine (1-3 per cent) (main active principle), aditerpene–16–O–methylcafestol, cafestol, kahweol, kuran–18–oic acid, glucopyranosyl ester-cofaryloside and atractyligenin glycoside.
Bilogical Activity	:	Respiratory stimulant activity found positive.
References	:	Bhattacharjee (1998), Rastogi and Mehrotra (1991, 1995), Sharma (2003).

Colebrookea oppositifolia Sm.

Syn.: *C. ternifolia* Roxb.

Family: Lamiaceae

H. Names	:	*H.*: Binda, Bindu, Pansra.
Distribution	:	More or less throughout India on low hills.
Part/s Used	:	Roots and Stem.
Utilization	:	2 tsp infusion of roots given, twice daily for 21 days, to treat asthma. Stem useful for cough.
Active Constituents	:	Active principles of root are triacontane, triacontanol, β-sitosterol, palmitic, stearic and oleic acid and 5, 6, 7 - trimethoxyflavone and 4′, 5, 6, 7- tetramethoxyflavone.
References	:	Jain (1984), Jha and Varma (1996), Kirtikar and Basu (1984), Rastogi and Mehrotra (1993)

Coleus amboinicus Lour.

Syn.: *C. aromaticus* Benth; *Plectranthus aromaticus* Roxb.

Family: Lamiaceae

Eng., Sans. and H. Names	:	*Eng.*: Country Borage, Indian Borage;

		Sans.: Pashanabhedi;
		H.: Pathorchur.
Distribution	:	Throughout India.
Part/s Used	:	Leaves.
Utilization	:	Decoction of leaves given for chronic cough, asthma and bronchitis.
Active Constituents	:	Cirsimaritin and β- sitosterol–β–D–glucoside, oleanolic, 2α, 3α–dihydroxyolean -12-en-28-oic, crategolic, pomolic, euscaphic, tormentic, ursolic and 2α, 3α -, 19α, 23–tetrahydroxyurs–12–en–28 -oic, salvigenin, 6–methoxy–genkwan, quercetin, chrysoeriol, luteolin, apigenin, eriodictyol and taxifolin isolated from leaves.
References	:	Ambasta (1986), Bhandary and Chandrashekhar (2002), Caius (2003), Kirtikar and Basu (1984), Rastogi and Mehrotra (1991), Watt (1972).

Commiphora wightii (Arn.) Bhandari

**Syn.: *C. mukul* (Arnott.) Bhandari; *Balsamodendron wightii* Arnott.;
B. roxburghii Stocks; *B. mukul* Hook. ex Stocks.**

Family: Burseraceae

Eng., Sans. and H. Names	:	*Eng.*: Gugul, Gum;
		Sans.: Deveshta, Durga, Jatala, Mahisaksa, Pura;
		H.: Gogil, Guggal, Guggul.
Distribution	:	Bellary, Mysore, Deccan, Khandesh, Kathiawar and Rajputana Desert.
Part/s Used	:	Bark and Gum–Resin.
Utilization	:	Inhalation of smoke from gum resin cures bronchitis, nasal catarrh, laryngitis and phthisis. Resin with candy given in whooping cough. Fresh decoction of plant taken orally to cure asthma. Powdered bark prescribed with water for relief from cough and cold.
Active Constituents	:	Gum resin yields gugglusterone, guggulsterol–VI and Z-guggusterol, 20α-hydroxy-4-pregnen-3-one, 20β-hydroxyl–4–pregnen–3–one, 16β–hydroxy–4,17 (20Z)–pregnadien–3–one, 16α–hydroxyl–4–pregnen–3–one, monocyclic diterpene-α-camphorene and cembrene.
References	:	Anonymous (1999), Bennet (1987), Chopra *et al.* (1956), Das (1997), Kirtikar and Basu (1984), Mishra and Dixit (1976), Rastogi and Mehrotra (1991, 1993), Shah (1982), Singh and Pandey (1998), Sinha and Sinha (2001).

Corchorus aestuans L.

Syn.: *C. acutangulus* auct. non Forsk.

Family: Tiliaceae

Distribution	:	Hotter parts of India.
Part/s Used	:	Roots.
Utilization	:	1 tsp powdered mixture comprising 10g of its roots, 5g rhizome of asparagus, 5g leaves of *Ocimum tenuiflorum* taken, thrice daily for 15 days, for curing asthma.
Active Constituents	:	Plant contains quercetin.
References	:	Asolkar *et al.* (1992), Chopra *et al.* (1956), Rao and Pullaiah (2001).

Cordia dichotoma Forst.

Syn.: *C. myxa* Sensu Cl.; *C. lowriana* Brandis; *C. obliqua* Willd.

Family: Boraginaceae

H. Name	:	*H.*: Lasora.
Distribution	:	Common throughout India; often planted.
Part/s Used	:	Leaves and Fruit.
Utilization	:	Decoction of leaves useful for cough and cold.
Active Constituents	:	β-sitosterol and α-linoleic, palmitic, linoleic and oleic acid identified in leaves.
References	:	Asolkar *et al.* (1992), Das and Agarwal (1991), Rastogi and Mehrotra (1995), Saxena and Vyas (1983), Singh and Maheshwari (1983).

Cordia wallichii G. Don

Syn.: *C. obliqua* Willd. var. *wallichii*.

Family: Ehretiaceae

Eng., Sans. and H. Names	:	*Eng.*: Large Sebesten;
		Sans.: Bahuvaraka, Bhutavriksha, Shataka;
		H.: Buralessura.
Distribution	:	Gujarat–N. Kanara, Deccan and W. Ghats.
Part/s Used	:	Fruits.
Utilization	:	Unripe fruit cures cough and bronchitis.
References	:	Ambasta (1986), Kirtikar and Basu (1984).

Costus speciosus (Koenig.) Sm.

Syn.:*Banksia speciosa* Koenig;
Costus speciosus var. *nepalensis* (Rosc.) Baker.

Family: Zingiberaceae

Sans. and H. Names	:	*Sans.*: Kashmira, Kustha, Padmakarna, Padmapunya, Pushkara, Sagara, Subandhu, vira.
		H.: Keu, Kust.
Distribution	:	More or less throughout India upto 1,500m.
Part/s Used	:	Rhizome.
Utilization	:	Rhizome considered effective in treating bronchitis and cough.
Active Constituents	:	Tetradecyl13-methylpentadecanoate tetradecyl 11-methyltridecanoate, 14-oxotricosanoic acid, 14-oxoheptacosanoic acid and 15-oxooctacosanoic, 31-norcycloartanone, cycloartanol, cycloartenol 7 cyclolaudenol, tigogenin, diosgenin, saponin A and B and β- sitosterol isolated from rhizome.
References	:	Bennet (1987), Bhattacharjee (1998), Jain and Tarafdar (1970), Kirtikar and Basu (1984), Rastogi and Mehrotra (1991), Sharma (2003).

Crinum defixum Ker-Gawl

Syn.: *Belutta polytaly* Rheede; *C. asiaticum* Roxb. (non Linn.);
C. roxburghii Dals.; *Radix toxicaria* Secunda;

Family: Amaryllidaceae

Distribution	:	Swampy river banks throughout India, Ceylon.
Part/s Used	:	Leaves
Utilization	:	1 tsp extract of leaves with equal quantity of honey given, thrice a day, to cure asthma.
References	:	Asolkar *et al.* (1992), Khanna and Kumar (2000), Kirtikar and Basu (1984), Watt (1972).

Crocus sativus Linn.

Family: Iridaceae

Eng., Sans. and H. Names	:	*Eng.*: Saffron Crocus, Spanish Saffron
		Sans.: Agnishikha, Asrika, Chandana, Dipaka, Jaguda, Kaleyaka, Kunkuma, Pitaka
		H.: Kesar, Zafran.
Distribution	:	Cultivated mainly in Kashmir upto 1,600m.

Part/s Used	:	Flower (Female part).
Utilization	:	Paste of stigma with cow milk taken, twice a day, for cough and cold. 10g of mixture comprising 60g 'kesar' (saffron) and 1tsp of honey kept in bottle for 7 days in a drum containing grains of *Triticum aestivicum*, taken on empty stomach for curing cough and cold.
Active Constituents	:	These are crocin, crocetin, picrocrocin (herb), carotene, lycopene, riboflavin and thiamine.astragalin, helichryysoside, kaempferol, kaempferol-3-O-β-D-glucosyl (1→2)-β-D-(6-acetyl) glucopyranoside and kaempferol-3-o-β-D-glucosyl (1→2)-β-D-glucopyranoside (petals).
Biological Activity	:	Extract of saffron has been found to stimulate respiration.
References	:	Kirtikar and Basu (1984), Rastogi and Mehrotra (1995), Sen and Batra (1997), Shome *et al.* (1996), Sharma (2003), Sinha and Sinha (2001).

Croton bonplandianum Baill.

Syn.: *C. sparsiflorus* Morong.

Family: Euphorbiaceae

Distribution	:	Throughout plains of India.
Part/s Used	:	Leaves, Stem and Inflorescence.
Utilization	:	Decoction of stem, leaves and inflorescence along with sugar given, twice a day, for checking cough. Also, inhalation of vapours of plant by boiling it in water considered good for breathing troubles.
Active Constituents	:	3- methoxy-4, 6- dihydroxymorphinandien-7-one (I) and norsinoacutine; alkaloid- sparsiflorine; rutin isolated from leaves. β- sitosterol and taraxerol, vomifoliol and ursolic acid isolated from both leaves and stem.
References	:	Bennet (1972), Rastogi and Mehrotra (1990, 1991, 1993, 1995), Verma *et al.* (1995).

Cupressus sempervirens Linn.

Family: Cupressaceae

Eng., Sans. and H. Names	:	*Eng.*: Mediterranean Cypress;
		Sans.: Surahva;
		H.: Sara, Saras, Saru.

Distribution	:	N. W. India; known in the cultivated state.
Part/s Used	:	Leaves.
Utilization	:	Leaves yield oil of cypress employed for whooping cough.
Active Constituents	:	Karahanaenone, quercetin and its 3-O-α-L- rhamnoside, hinokiflavone and isocryptomerin and α- pinene isolated from leaves.
References	:	Ambasta (1986), Kirtikar and Basu (1984), Rastogi and Mehrotra (1995).

Curculigo orchioides Gaertn.

Syn.: *C. malabarica* Wight.;*Hypoxis orchioides* Kurz.

Family: Hypoxidaceae

Sans. and H. Names	:	*Sans.*: Ashoghni, Godhapadi, Khalani, Mushali, Suvaha, Tali;
		H.: Kali musli, Mushali, Muslikand, Siyahmusli.
Distribution	:	Distributed in Peninsular India. Khasia Hills. Sub–tropical Himalayas.
Part/s Used	:	Tuber.
Utilization	:	Half tsp of powdered tuber mixed with honey given, twice a day, for 3 days to cure cough and bronchitis. Tuberous roots also considered good for asthma.
Active Constituents	:	These are 2 methoxy–4–acetyl–5–methyl triacontane. 2 aliphatic hydroxy ketones–27-hydroxytriacontan-6-one and 23-hydroxy-triacontan-2-one, orcinol glycoside– corchioside A (root), curculigol, 6 triterpenoid saponins-curculigo saponins A,B,C,D,E and F bearing common genin- curculigenein A, hentriacontanol, sitosterol, stigasterol, cycloartenol and sucrose (rhizome).
References	:	Ambasta (1986), Chaudhary and Hutke (2003), Kirtikar and Basu (1984) Prasad and Abraham (1984), Rao and Henry (1996), Rastogi and Mehrotra (1993, 1995), Saxena and Dutta (1975), Sharma (2003), Sinha and Sinha (2001), Watt (1972).

Curcuma amada Roxb.

Family: Zingiberaceae.

Eng., Sans. and H. Names	:	*Eng.*: Mango Ginger;
		Sans.: Karpura- Haridra, Amragandha, Karpura, Padmapatra, Suranayika;
		H.: Ama haldi, Kapurahaladi.

Distribution	:	Bengal. W. Peninsula.
Part/s Used	:	Rhizome.
Utilization	:	Decoction of rhizome (5ml) considered useful for asthma and bronchitis, twice a day.
Active Constituents	:	Main constituents of rhizome are d-α-pinene, d-camphor, 1, β–curcumene as hydrochloride, d-curcumene and phytosterol.
References	:	Kirtikar and Basu (1984), Rastogi and Mehrotra (1990), Tripathi and Goel (2001).

Curcuma angustifolia Roxb.

Family: Zingiberaceae

Eng., Sans. and H. Names	:	*Eng.*: Arrow root, East India, Narrow leaved turmeric, Wild Arrow root;
		Sans.: Godhumaja, Payakshira,Tavakshir, Talakshira, Yavaja;
		H.: Tavakhira, Tikhur.
Distribution	:	Outer ranges of C. Himalaya, W. Bihar, N. Bengal extending to Mumbai and S. India.
Part/s Used	:	Rhizome.
Utilization	:	10 ml decoction of rhizome given, twice a day, for asthma. Also, decoction of 1tsp of its powder along with 2 tsp seeds of *Trachyspermum ammi* useful for cough and cold.
Active Constituents	:	α- and β- pinenes, (-) ar-cucumene, (+) camphor, (+) α-terpineol, (+) borneol and zingiberol isolated from rhizome.
References	:	Kirtikar and Basu (1984), Panthi and Chaudhary (2003), Rastogi and Mehrotra (1993), Sharma (2003), Tripathi and Goel (2001).

Curcuma longa Linn.

Syn.: *C. domestica* Valeton

Family: Zingiberaceae

Eng., Sans. and H. Names	:	*Eng.*: Indian Saffron,Turmeric;
		Sans.: Gharshani, Haridra, Hemaragini, Kanchani, Mangalaya, Pavitra, Rajani, Ratri, Shifa;
		H.: Haldi, Haridra.

Distribution	:	Cultivated in the plains of India.
Part/s Used	:	Leaves, Roots and Rhizome.
Utilization	:	Decoction of leaves given, thrice daily for 3-5 days, for cold and pneumonia. Green rhizome given for whooping cough and bronchitis. "Haridra khand", a compound containing powdered turmeric and sugar is a known preparation taken for cough and cold. Powdered roots considered useful for bronchitis. Paste made of rhizome powder, molasses and mustard oil when chewed gives relief from problems of respiratory tracts.
Active Constituents	:	Rhizome contains antioxidant compounds-curcumin, 4-hydroxy cinnamoyl (feruloyl) methane and bis-(4-hydroxy cinnamoyl) methane, dihydrocurcumin, diferuloylmethane, feruloyl-p-coumaroylmethane and di-p-coumaroylmethane.
References	:	Anonymous (1999), Bhatt *et al.* (2001), Jain *et al.* (1997), Rastogi and Mehrotra (1993), Sen and Batra (1997), Shome *et al.* (1996), Thakur *et al.* (1989), Yusuf *et al.* (2002).

Curcuma zedoria (Christ.) Rosc.

Syn.: *C. zerumbet* Roxb.; *Amomum zedoria* Christm.

Family: Zingiberaceae

Eng., Sans. and H. Names	:	*Eng.*: Zedoary;
		Sans.: Dravida, Durlabha, Gandharara, Jatala;
		H.: Kachora, Kalihaladi.
Distribution	:	Grows wild in the E. Himalaya and in moist deciduous forest of coastal tract of Kanara.
Part/s Used	:	Rhizome.
Utilization	:	Decoction of rhizome along with long pepper, cinnamon and honey given for cold, flu and asthma.
Active Constituents	:	Rhizome yields essential oil containing α-pinene, camphor, sesquiterpenes, guaiane- zedoarondiol, sesquiterpenoids–(-) curcumenone, curcumanolides A and B, curcumol, zederone and furanodiene.
References	:	Kirtikar and Basu (1984), Rastogi and Mehrotra (1990, 1995), Tripathi and Goel (2001), Thakur *et al.* (1989).

Cuscuta reflexa Roxb.

**Syn.: *C. grandiflora* Wall.; *C. verucosa* Sweet.; *C. marantha* Don.;
C. santapani Banerji and Das.**

Family: Cuscutaceae

Eng., Sans. and H. Names	:	*Eng.*: Dodder;
		Sans.: Akashavalli, Amaravela, Nilatar;
		H.: Akashbela, Amarabela.
Distribution	:	Throughout India; grows upto 2,800m in various parts of H.P.
Part/s Used	:	Whole Plant.
Utilization	:	Decoction of plant useful for asthma, and in combination with black pepper, *Ocimum sanctum* and butter used, thrice a day for 15 days, in cough and bronchitis.
Active Constituents	:	Plant contains amerbelin, cussutin, wax, scoparone, melanettin, quercetin and hyperoside.
References	:	Kirtikar and Basu (1984), Rastogi and Mehrotra (1995), Sharma (2003), Singh (1996), Singh and Pandey (1998), Watt (1972).

Cydonia vulgaris Pers.

Syn.: *Pyrus cydonia* Linn.; *Cydonia oblonga* Mill.

Family: Rosaceae

Eng., Sans. and H. Names	:	*Eng.*: Quince tree;
		Sans.: Amrutphala;
		H.: Abi, Bihi.
Distribution	:	Cultivated in N.W. India.
Part/s Used	:	Fruits and Seeds.
Utilization	:	Fruits considered useful in asthma and cough.
Active Constituents	:	Marmelolactones A and B, glucosides of procyanidin polymers, flavonols, hyperin, isoquercitrin, quercetin and terpenoid lactones isolated from fruits.
References	:	Asolkar *et al.* (1992), Kirtikar and Basu (1984), Rastogi and Mehotra (1993, 1995), Watt (1972).

Cyperus scariosus R.Br.

Syn.: *Cyperus pertenuis* Roxb.

Family: Cyperaceae

Sans. and H. Names	:	*Sans.*: Chakranksha, Charukesara, Nargusta, Shishira, Uchchta;
		H.: Nagarmutha.
Distribution	:	Damp places in W. Bengal, U.P. Eastern and Southern parts of India, Kutch and Saurashtra.
Part/s Used	:	Whole Plant.
Utilization	:	10 ml decoction of whole plant with 5 g *Piper nigrum* given, twice or thrice daily for 4-5 days, to cure cough and cold.
References	:	Ambasta (1986), Kirtikar and Basu (1984), Maliya and Singh (2003), Thakur *et al.* (1989).

Cymbopogon citratus (DC.) Stapf.

Syn.: *Andropogon citratus* DC.

Family: Poaceae

Eng., Sans. and H. Names	:	*Eng.*: Melissa grass, West Indian lemon Grass;
		Sans.: Badhira, Bhustrina, Bhutika;
		H.: Gandhatrina.
Distribution	:	Warm humid areas.
Part/s Used	:	Leaves.
Utilization	:	Infusion of leaves considered useful against cough, cold and bronchitis.
Active Constituents	:	Leaves contain luteolin, its 7-O-glucoside, 7-O-neohesperoside, homo orientin and its 2″-O-rhamnosyl derivatives.
References	:	Ambasta (1986), Caius (2003), Gill *et al.* (1997), Kirtikar and Basu (1984), Rastogi and Mehrotra (1995), Sharma and Sood (1997).

Dalbergia sexatilis Hook.F.

Family: Fabaceae

Part/s Used	:	Leaves, Bark and Roots.
Utilization	:	Decoction of leaves, bark and roots prescribed for bronchial ailments.

Active Constituents	:	These contain tannin and essential oil.
Reference	:	Gill and Nyawuame (1994).

Datura innoxia Mill.

Syn.: *D.metel* sensu Cl.

Family: Solanaceae

Distribution	:	Temperate Himalaya upto 2,500m and in hilly regions of C. and S. India.
Part/s Used	:	Leaves, Flowers and Seeds.
Utilization	:	Leaves smoked to cure asthma. Extract of seeds also useful for asthma. Powdered dried leaves mixed with powder of black pepper and honey or jaggery for alleviating constant coughing.
Active Constituents	:	These are hyosyamine, scopolamine, meteloidine (leaves and seeds), quercetin–7- glucosido–3 - sophoroside, quercetin–7–glucosido–3 -glucogalacto-side and their esters with caffeic and p-coumaric acid; Kaemferol-7-glucosido-3-sophoroside, small amounts of hyoscine and atropine (leaves).
References	:	Bennet (1987), Bhatt *et al.* (2001), Kshirsagar *et al.* (2003), Rastogi and Mehrotra (1990, 1991), Singh and Pandey (1998), Sharma (2003).

Datura metel Linn.

Syn.: *D.fastuosa* L.; *D.fastuosa* var.*alba* (Nees) Cl.

Family: Solanaceae

Eng., Sans. and H. Names	:	Eng.: Downy Datura;
		Sans.: Dhushtura;
		H.: Sadah-dhatura.
Distribution	:	Weed of wasteland all over India.
Part/s Used	:	Leaves, Flowers, Seeds and Roots.
Utilization	:	Leaves, flowers, seeds and roots considered good in spasmodic asthma. Leaves also smoked to cure cough.
Active Constituents	:	Leaves afford hyoscyamine (22-25 per cent), hyoscine A datumetine, atropine, scopolamine, daturilin, datumelin, secowithametelin, withametelin B, ithafastuosins A and B along with withametelin, daturilinol, datumetelin, withametelin, isowitha-metelin. Flowers yield withanolides-withametelin F and G.

References	:	Balasubramaniam and Prasad (1996), Bennet (1987), Nayak *et al.* (2003), Rastogi and Mehrotra (1991, 1995), Sharma (2003).

Datura stramonium Linn.

Syn.: *D.ferox* Nees; *D.wallichii* Dunal; *Stramonium vulgatus* Cl.; *D.stramonium* var.*tatula* (L.) Cl.

Family: Solanaceae

Eng. and Sans. Names	:	*Eng.*: Apple of Peru, Mad apple, Thorn apple
		Sans.: Dhatoora, Dhurta, Shivapriya
Distribution	:	Grown in Kashmir valley, Hills of H.P. and U.P.
Part/s Used	:	Leaves, Flowers, Roots and Seeds.
Utilization	:	Dried leaves and flowers smoked to alleviate asthma and bronchtis. Cloves kept for 24 hours in place of seeds removed from its fruits and given after roasting for checking asthmatic attacks.
References	:	Ambasta (1986), Bhattacharjee (1998), Kirtikar and Basu (1984), Rastogi and Mehrotra (1990, 1995) Sharma (2003), Singh (1999), Sinha and Sinha (2001), Upadhye *et al.* (1994), Watt (1972).

Delphinium cashmerianum Royle

Syn.: *D.aitchosonii* Huth.

Family: Ranunculaceae

Distribution	:	W. Himalaya.
Part/s Used	:	Leaves.
Utilization	:	Leaf decoction recommended for cough.
Active Constituents	:	Leaves afford ashmiradelphine, anthranoylycoctonine, lyocomitine, avadhirine, lappaconitine and its N–de–o Ac.
References	:	Asolkar *et al.* (1992), Bennet (1987), Chandrasekar and Srivastava (2003), Watt (1972).

Dendrophthoe falcata (Linn.f.)Ettingshausen

Syn.: *Loranthus falcatus* Linn.f.; *L.longiflorus* Desr.

Family: Loranthaceae

Sans. and H. Names	:	*Sans.*: Vanda;
		H.: Banda.

Distribution	:	More or less throughout India.
Part/s Used	:	Bark.
Utilization	:	Bark useful in asthma.
Active Constituents	:	(+) catechin and leucocyanidin isolated from bark.
References	:	Ambasta (1986), Chopra *et al.* (1956), Rastogi and Mehrotra (1993, 1995).

Descurainia sophia (Linn.) Web.

Syn.: *Sisymbrium sophia* Linn.

Family: Brassicaceae

Eng. and H. Names:	:	*Eng.*: Flixweed, Fluxweed;
		H.: Khukallana.
Distribution	:	Kashmir to Kumaon upto 4,650m.
Part/s Used	:	Whole Plant and Seeds.
Utilization	:	Infusion of plant with equal quantity of honey or vinegar recommended for chronic cough and asthma. Seeds also prescribed in bronchitis.
Active Constituents	:	Strophanthidin, evomonoside, helveticoside, evoioside and erysimoside isolated from seeds.
References	:	Caius (2003), Das and Agarwal (1991), Kapur *et al.* (1996), Rastogi and Mehrotra (1993).

Desmodium adscendens (Sw.) DC.

Syn.: *Hedysarum adscendens* Sw.; *D. thwaitesii* Baker

Family: Fabaceae

Part/s Used	:	Leaves.
Utilization	:	Powdered dried leaves given with warm water for bronchial asthma.
Active Constituents	:	Tannins.
References	:	Bennet (1987), Gill and Nyawuame (1994).

Desmodium gangeticum (L.) DC.

Syn.: *Hedysarum gangeticum* Linn., *D. collinum* Roxb.; *D. gangeticum* var. *maculatum* (L.) Baker

Family: Fabaceae

Sans. and H. Names	:	*Sans.*: Anshumati, Dhruva, Ekamula, Shalani;
		H.: Salpani, Salun, Salwan.

Distribution	:	Outer Himalaya upto 1,700m and throughout India.
Part/s Used	:	Roots and Whole Plant.
Utilization	:	Decoction of roots and the plant prescribed for asthma, cough and cold.
Biological Activity	:	Mixture of root alkaloid has been found to increase rate and amplitude of respiration in dog.
References	:	Das (1997), Gill and Nyawuane (1994), Kirtikar and Basu (1984), Sinha and Sinha (2001), Sinha *et al.* (2002), Thakur *et al.* (1989).

Desmodium triquetrum (L.) DC.
Family: Fabaceae

Distribution	:	C. and Eastern Himalaya ascending to 1,350m in Kumaon, Sikkim and Khasia Hills to S. India.
Part/s Used	:	Leaves and Roots.
Utilization	:	Paste of leaves made along with pepper and jaggery taken, twice a day, for asthma. Decoction of leaves and roots also given to cure cough and cold.
Active Constituents	:	Friedelin, epifriedelinol, and stigasterol isolated from the leaves.
References	:	Bhandary *et al.* (1996), Chopra *et al.* (1980), Kumar (2002), Rastogi and Mehrotra (1995).

Desmos longiflorus (Roxb.) Safford
Family: Annonaceae

Distribution	:	Assam and Meghalaya upto 900m.
Part/s Used	:	Roots.
Utilization	:	Infusion of roots (10ml), twice daily till cure, considered good by the Bawn tribe for asthma.
References	:	Bennet (1987), Lalramnghinglova and Jha (1997), Rastogi and Mehrotra (1995).

Dicranopteris linearis (Burm.) Underwood
Syn.: *Gleichenia linearis* Bedd.; *G.linearis* C.B.Clarke; *G. dichotoma* Willd.
Family: Gleicheniaceae

Distribution	:	Throughout India ascending upto 2,000m.
Part/s Used	:	Fronds and Underwood Plant.

Utilization	:	Fronds and underwood regions of plant recommended for cure of asthma.
References	:	Chopra *et al.* (1956), Singh (1999).

Digitalis purpurea Linn.

Family: Scrophulariaceae

Eng. Names	:	*Eng.*: Foxglove, Purple Foxglove.
Distribution	:	Cultivated in Kashmir and H.P as an ornamental plant: 1,500-1,800m.
Part/s Used	:	Leaves.
Utilization	:	Leaf infusion given for relief from asthma.
Active Constituents	:	Principle constituents are digitoxin, gitoxin, gitalin, tannins, inositol and luteolin.
References	:	Bhattacharjee (2001), Chopra *et al.* (1956), Rastogi and Mehrotra (1990, 1995), Sharma (2003), Sinha and Sinha (2001), Thakur *et al.* (1989).

Dillenia indica Linn.

Syn.: *D. speciosa* Thunb.; *D. elliptica* Thunb.

Family: Dilleniaceae

Sans. and H. Names	:	*Sans.*: Bhavya;
		H.: Chalta.
Distribution	:	Evergreen forests of Sub-Himalayan tract from Kumaon to Assam, Bengal and S.wards to C. and S. India.
Part/s Used	:	Fruits.
Utilization	:	Fruit juice mixed with sugar and water used as cough mixture.
Active Constituents	:	Neutral polysaccharides containing arabinose (20 per cent), galactose (55 per cent) and mannose (18 per cent) isolated from mucilaginous core of fruits.
References	:	Das and Agarwal (1991), Kirtikar and Basu (1984), Kumar (2002), Rastogi and Mehrotra (1993), Watt (1972).

Dioscorea bulbifera Linn.

Syn.: *D. longifolia* Benth.; *D.sativa* auct. non L.; *D. pulchella* Roxb.; *D. tamnifolia* Salib.; *D. pulchella* Hokenhacker; *Helmia bulbifera* Kunth.

Family: Dioscoreaceae

Eng., Sans. and H. Names	:	*Eng.*: Bulb bearing yam;

Sans.: Brahmaputri, Brahmikanda, Ghrishti, Varahi;

H.: Genth, Genthi, Karukanda.

Distribution	:	Throughout India ascending upto 2,000m in Himalaya.
Part/s Used	:	Tubers.
Utilization	:	Tubers considered useful for asthma and bronchitis.
Active Constituents	:	Furanoid norditerpenes–diosbulinoside D and F, a dihydrophenanthrene, phenanthrene, diosbulbins B and D and 5-sorbitol isolated.
References	:	Chopra *et al.* (1956), Kirtikar and Basu (1984), Rastogi and Mehrotra (1993, 1995), Sharma (2003).

Diospyros kaki Linn.

Syn.: *D. chinensis* Bl.; *Embryoptris kaki* Don; *D. costata* Rev.Hort.

Family: Ebnaceae

Distribution	:	Khasia Hills and Upper Assam.
Part/s Used	:	Calyx and Fruit Peduncle.
Utilization	:	Recommended for Cough.
Active Constituents	:	Lignan-(-) dvanillyltetrahydrofuran isolated from fruits.
References	:	Das and Agarwal (1991), Watt (1972).

Diospyros melanoxylon Roxb.

Syn.: *D. wightiana* Wall; *D. dubia* Wall.

Family: Ebenaceae

Distribution	:	India. Malabar. Orissa.
Part/s Used	:	Stem-bark.
Utilization	:	Stem-bark extract mixed with jaggery (1tsp, twice a day, till cure) prescribed for cough and cold by Konda tribes.
Active Constituents	:	Ceryl alcohol, lupeol, betulin, β-sitosterol, sequoyitol and triterpene carboxylic acid-diospyric acid isolated rom Stem-bark.
References	:	Rao and Henry (1996), Rastogi and Mehrotra (1990), Watt (1972).

Diplazium esculentum (Retz.) Sw.

Syn.: *Anisogonium esculentum* (Retz.) Presl.

Family: Pteridaceae

Distribution	:	Throughout the plains of India, upto 1,000 m.

Part/s Used	:	Rhizome.
Utilization	:	Recommended for cough and asthma.
Active Constituents	:	Rhizome affords protocatechuic and syringic acid.
References	:	Rastogi and Mehrotra (1993, 1995), Singh (1999).

Drosera peltata Sm.
Syn.: *D. lunata* Buch-Ham.
Family: Droseraceae

H. Name	:	*H.*: Mukhajali.
Distribution	:	Throughout the Himalaya.
Part/s Used	:	Herb.
Utilization	:	Dried herb used in chronic bronchitis, asthma and whooping cough.
Active Constituents	:	These are resin, plumagin and a proteolytic enzyme of pepsin type.
Biological Activity	:	Extract of leaves and stem found to inhibit growth of *Mycobacterium tuberculosis*.
References	:	Asolkar *et al.* (1992), Rastogi and Mehrotra (1990, 1995), Sharma (2003), Watt (1972).

Drymaria cordata (L.) Willd. *ex* Roem. and Schult.
Syn.: *D. diandra* Bl.; *D. villosa* Cham and Sch.; *D. villosa* ssp. *villosa* Sensu Mizushima; *D. cordata* ssp. *diandra* (Bl.) Duke; *Holosteum cordatum* L.
Family: Caryophyllaceae

Part/s Used	:	Whole Plant.
Utilization	:	Shoots recommended for curing asthma and the whole plant prescribed for cough, cold and pneumonia.
Active Constituents	:	Palmitic, succinic, p-hydroxycinnamic acid and α spinasterol isolated from plant.
References	:	Bennet (1972), Jain and Saklani (1992).

Eclipta alba (L.) Hassk.
Syn.: *E. prostrata* Hook.; *E. erecta* Linn.; *E. prostrata* L.; *Verbesina prostrata* L.
Family: Asteraceae

Sans. and H. Names	:	*Sans.*: Ajagra, Bhringa, Keshya, Karanjaka;
		H.: Babri, Bhangra, Mockand.

Distribution	:	Abundant throughout India; ascending to 2,000m on the Himalaya.
Part/s Used	:	Leaves.
Utilization	:	Fresh decoction of leaves (a glassful) taken with a little honey diluted in water, thrice a day, for cough and cold.
Active Constituents	:	Leaves contain α–terthienyl–methanol, stigasterol, α–amyrin mixture of desmethyl wedeloacetone.
References	:	Anonymous (1999), Caius (2003), Kirtikar and Basu (1984), Rastogi and Mehrotra (1993), Singh *et al.* (1991), Sharma (2003), Watt (1972).

Elaeagnus angustifolia Linn.

Syn.: *E.hortensis* Bieb.; *E. orientalis* Linn.

Family: Elaeagnaceae

H. Name	:	H.: Shiulick.
Distribution	:	W. Himalaya: 1,700 -3,700m.
Part/s Used	:	Seeds.
Utilization	:	Seed oil considered good in catarrhal and bronchial affections.
References	:	Ambasta (1986), Kirtikar and Basu (1984), Watt (1972).

Elaeagnus umbellata Thunb.

Syn.: *E. parvifolia* Wall.

Family: Elaeagnaceae

H. Name:		H.: Ghain.
Distribution	:	Temperate Himalaya from Kashmir to Nepal 915-3,048m.
Part/s Used	:	Seeds and Flowers.
Utilization	:	Seeds and flowers used as stimulant in cough and pulmonary affections.
References	:	Ambasta (1986), Das and Agarwal (1991), Watt (1972).

Elephantopus scaber Linn.

Family: Asteraceae

Eng. Name	:	*Eng.:* Prickly leaved elephant's foot.
Distribution	:	Distributed throughout the hotter parts of India.

Part/s Used	:	Roots.
Utilization	:	Pounded roots made into solution and given for cough.
Active Constituents	:	Deacyl cynaropicric, glucozaluzanin C, crepiside E and stigasteryl-3-β-glucoside isolated from roots.
References	:	Jain (1968), Jain *et al.* (1997), Rastogi and Mehrotra (1995), Watt (1972).

Elsholtzia blanda Benth.
Syn.: *Aphanochilus blandus* Benth.
Family: Lamiaceae

Distribution	:	Himalayas. Meghalaya. Hills of Bihar.
Part/s Used	:	Leaves.
Utilization	:	Poultice of leaves applied on chest of a child for allaying cough and cold.
References	:	Manandhar (1990), Rastogi and Mehrotra (1995).

Elytraria acaulis (L.f.) Lindau
Syn.: *Justicia acaulia* L.; *E. crenata* Vahl.
Family: Acanthaceae

Part/s Used	:	Leaves
Utilization	:	Dried leaves inhaled in bronchitis.
References	:	Bennet (1987), Painuli and Maheshwari (1996).

Emblica officinalis Gaertn.
[Pl. 2C]
Syn.: *Phyllanthus emblica* Linn.
Family: Euphorbiaceae

Sans. and H. Names	:	*Sans.*: Amlika;
		H.: Aonla.
Distribution	:	Throughout Tropical and Sub-Tropical India.
Part/s Used	:	Seeds and Fruits.
Utilization	:	Decoction of powdered seeds used against asthma and bronchitis. Powdered dried fruits along with that of *Beleria myrobalan* (in equal proportion), given, 3 times a day, to check cold and flu; also eaten raw to cure cough.
Active Constituents	:	Trigalloylglucose, terchebin, corilagin, ellagic acid, IAA and 4 auxins detected in mature fruits. Seeds contain fixed oil, phosphatides and essential oil.

References	:	Aminuddin and Girach (1991), A n o n y m o u s (1999), Bhattacharjee (2001), Caius (2003), Das and Agarwal (1991), Jain (1968), Kothari and Rao (1999), Mandal and Basu (1996), Rastogi and Mehrotra (1990, 1993) Sharma (2003), Sinha and Pandey (1998).

Embelia tsjeriam–cottam (Roemer and Schult) DC.

Syn.: *E. robusta* non Roxb.

Family: Myrsinaceae

H. Names	:	*H.*: Baibirang, Bhungi.
Distribution	:	Throughout the greater parts of India upto 1,700m.
Part/s Used	:	Fruits.
Utilization	:	Decoction of powdered dried fruits good against tuberculosis.
Active Constituents	:	Dried berries yield 1.6 per cent embelin, 0.15 per cent potassium hydrogen oxalate and 7.5 per cent fatty ingredients.
Biological Activity	:	Respiratory stimulation and antitubercular activities confirmed.
References	:	Bhandary and Chandrashekhar (2002), Chopra *et al.* (1956), Jain (1968) Sharma (2003), Thakur *et al.* (1992).

Ephedra gerardiana Wall. ex Stapf.

Syn.: *E. vulgaris* Hook. f.

Family: Ephedraceae

H. Name	:	*H.*: Asmania.
Distribution	:	Drier regions of higher Himalaya: 2,000-4,000m.
Part/s Used	:	Leaves, Roots and Berries.
Utilization	:	Powdered plant (2-3g) prescribed, thrice daily for 3-5 days, for cough and asthmatic attacks. Juice of berries used in respiratory troubles. Dried stem collected in autumn, constitute the drug ephedra, used in treatment of bronchial asthma.
Active Constituents	:	These are benzylmethylamine and 2, 3, 4–trimethyl–5 - phenyl oxazolidine (aerial parts of plant), 1-tyrosine betaine–manokonine (roots), ephedrine (0.3 per cent) and pseudoephedrine (stem).
References	:	Ambasta (1986), Bhattacharjee (2001), Chaurasia *et al.* (1999), Jain (1968), Kaul (1997), Misra and Kumar

(2000), Raju (2000), Rao (1996), Sharma (2003), Sharma and Rana (1999), Sharma and Sood (1997), Sood *et al.* (2001), Thakur *et al.* (1989).

Epimedium elatum Morr. and Decne.

Family: Podophyllaceae

Part/s Used	:	Leaves.
Utilization	:	Extract of leaves useful for asthmatic paraxysms.
Reference	:	Kapur (1996).

Erioglossum rubiginosum Blume

Syn.: *E. edule* Blume; *Sapindus rubiginosa* Bl.

Family: Sapindaceae

H. Name	:	*H.*: Ritha.
Distribution	:	N.E. India, Deccan Peninsula and Andaman and Nicobar Islands.
Part/s Used	:	Seeds and Shoot.
Utilization	:	Tender shoots given for cough and decoction of seeds for whooping cough.
References	:	Ambasta (1986), Das and Agarwal (1991), Watt (1972).

Erycibe paniculata Roxb.

Syn.: *E. paniculata.* var. *wightiana* (Grah.) Cl.

Family: Convolvulaceae

Distribution	:	Throughout India.
Part/s Used	:	Leaves, Twigs and Bark.
Utilization	:	Decoction prepared by taking equal quantity of leaves, twigs and bark given for bronchial asthma.
References	:	Bennet (1987), Brahmam and Saxena (1990), Kirtikar and Basu (1984).

Erythrina stricta Roxb.

Family: Fabaceae.

Sans. Name	:	*Sans.*: Mura.
Distribution	:	W. half of Peninsula.
Part/s Used	:	Bark.
Utilization	:	Pulverized bark effective in treating asthma.

Active Constituents	:	7–methoxy- 8-(15-hydroxypentadecyl) coumarin, erysovin, erysodine, sitosterol, hypaphorine and stigasterol isolated from bark.
References	:	Ambasta (1986), Bhatacharjee (2001), Rastogi and Mehrotra (1993), Watt (1972).

Erythrophloeum suaveolens (Guill. and Pers.) Breman.
Family: Fabaceae

Part/s Used	:	Bark.
Utilization	:	Infusion of bark useful for cold.
Active Constituents	:	Cassamidine, erythrophlamide and erythrophleine isolated from bark.
Reference	:	Gill and Nyawuame (1994).

Erythroxylon coca Lam.
Family: Erythroxylaceae

Eng. Names	:	*Eng.*: Bolivian coca, Coca bush, Cocaine.
Distribution	:	Cultivated in various parts of India.
Part/s Used	:	Leaves.
Utilization	:	Infusion of leaves used for asthma. Also, leaves smoked or chewed to relieve asthma.
Active Constituents	:	Leaves contain 0.7-2.5 per cent total alkaloids- cocaine, cinnamylescaine and β-truxilline coco, cocatannic acid, (-) dihydrocuscohygrine, meso–and (-) cuscohygrine and cis- and trans- cinnamoyl cocains.
References	:	Bhattacharjee (2001), Greenish (1999), Kirtikar and Basu (1984), Rastogi and Mehrotra (1993), Watt (1972).

Eucalyptus globulus Labill.
Family: Myrtaceae

Eng. Names	:	*Eng.*: Blue Gum Tree of Victoria.
Distribution	:	H.P., Kumaon hills, Darjeeling, Shillong and S. Indian hills.
Part/s Used	:	Leaves.
Utilization	:	Leaves smoked for allaying asthma. Oil from leaves useful in bronchial cases and in infection of upper respiratory tract.
Active Constituents	:	3-5 per cent of volatile oil, cis and trans- pinocarveol, α-pinene, an unsaturated, α-ketones, oleanolic acid, maslinic acid, gallic, caffeic, ferulic, protocatechuic acid,

quercitol, quercitrin, rutin, hyperoside and quercitol-3-glucoside isolated from leaves.

References	:	Chopra *et al.* (1956), Greenish (1999), Kirtikar and Basu (1984), Nunez and Castro (1995), Rastogi and Mehrotra (1990, 1991, 1995), Sharma (2003), Watt (1972).

Eupatorium odoratum Linn.

Syn.: *Chromolaena odorata* (L.) King and Robinson.

Family: Asteraceae

Distribution	:	A weed in Bengal and Assam.
Part/s Used	:	Flowers and Leaves.
Utilization	:	Extract or infusion prescribed for cough and cold.
Active Constituents	:	Ceryl alcohol, α-, β- and γ- sitosterol, isokuranatin, acacetin and a new chalcone odoratin isolated from leaves
References	:	Chopra *et al.* (1956), Gill *et al.* (1997), Rastogi and Mehrotra (1990, 1991, 1993, 1995).

Euphorbia hirta Linn.

Syn.:*E. pilulifera* auct. non Linn.; *E.pilulifera* L.; *Chamaesvce hirta* (L.) Millsp.

Family: Euphorbiaceae

Eng., Sans. and H. Names	:	*Eng.*: Australian asthma herb, Snake weed;
		Sans.: Nagarjun, Pusitoa;
		H.: Dudhi.
Distribution	:	Throughout India; common in warmer regions.
Part/s Used	:	Whole Plant and Leaves.
Utilization	:	Extract of leaves given, twice a day for 2 days, to cure pneumonia. Decoction of plant taken orally to cure asthma and bronchial infections.
Active Constituents	:	Saponin and tannin, 2 dehydroellagitannins-euphorbins A and B; quercetin, kaempferol, afzelin, myricetin, quercitrin, myricitrin, quercetin-7-glucoside, myo-inositol, gallic acid and protocatechuic acid isolated from whole plant.
Biological Activity	:	Antitubercular activity of plant confirmed.
References	:	Ambasta (1986), Bhatt *et al.* (1999), Bhatacharjee (2001), Caius (2003), Gill *et al.* (1997), Kirtikar and Basu (1984), Mishra (2003), Rastogi and Mehrortra (1995), Singh and Pandey (1998), Swami and Gupta (1996).

Euphorbia hixtra Linn.

Family: Euphorbiaceae

Part/s Used	:	Whole Plant.
Utilization	:	Plant extarct useful for cough and asthma.
Reference	:	Kumar (2002).

Euphorbia neriifolia Linn.

Syn.: *E.ligularia* Roxb., *E. neriifolia* Sensu Hook. f. non L.

Family: Euphorbiaceae

Sans. and H. Names	:	*Sans.*: Patrasnuchi, Svarasana;
		H.: Sehund, Sij, Thoar.
Distribution	:	W. Peninsula, cultivated elsewhere.
Part/s Used	:	Whole plant
Utilization	:	Extract of plant given, 3 times a day, for asthma.
Active Constituents	:	Neriifolene along with antiquorin isolated.
References	:	Chopra *et al.* (1956), Kirtikar and Basu (1984), Rastogi and Mehrortra (1995).

Euphorbia tirucalli Linn.

Family: Euphorbiaceae

Eng., Sans. and H. Names	:	*Eng.*: India tree, Milk bush, Spurge;
		Sans.: Bahukshira, Ganderi, Snuka, Trikuntaka;
		H.: Sehud, Sehund, Sindh, Thora.
Distribution	:	Naturalized in drier parts of Bengal, Deccan and S. India.
Part/s Used	:	Whole Plant.
Utilization	:	Plant extract useful in whooping cough and asthma.
Active Constituents	:	Hentriacontane, hentriacantrol, β-sitosterol, taraxerol, 3, 3-di-O-methylellagic acid and ellagic acid isolated from stem.
References	:	Caius (2003), Kirtikar and Basu (1984), Rastogi and Mehrotra (1991, 1990).

Evolvulus alsinoides Linn.

Syn.: *E.hirsutus* Lam.; *E. angustifolius* Roxb.

Family: Convolvulaceae

Sans. and H. Names	:	*Sans.*: Laghuvishnukranta;
		H.: Shyamakranta.

Distribution	:	Nearly all over India.
Part/s Used	:	Whole Plant.
Utilization	:	Leaves rolled into cigarette and smoked in chronic bronchitis and asthma. Decoction of the plant also useful in asthma and bronchitis.
Active Constituents	:	Plant contains evolvine and optically inactive fatty residue.
References	:	Bhatt *et al.* (2002), Chopra *et al.* (1980), Kirtikar and Basu (1984), Rastogi and Mehrotra (1991), Watt (1972).

Fagonia cretica Linn.

Syn.: *F. arabica* Linn.; *F. bruguieri* DC.; *F. mysorensis* Roth.

Family: Zygophyllaceae

Sans. and H. Names	:	*Sans.*: Ajabhakshya, Ananto, Dhanvi, Dusparsha;
		H.: Hinguna, Ustarkhar, Usturgar.
Distribution	:	Throughout N. W. India, Punjab and Southern regions of W. Peninsula.
Part/s Used	:	Whole Plant.
Utilization	:	Decoction of plant given for asthmatic bronchitis.
Active Constituents	:	These are triterpenoid saponins and characterized as 23, 28–di–O–β–D -glucopyranosyl–taraxer–20–en–28–oic acid and 3β,28–di–O–β–D–glucopyranosyl–23 hydroxytaraxer–20–en–28–oic acid (aerial parts), quercetin and kaemferol (leaves and flowers).
References	:	Ambasta (1986), Bhatt *et al.* (2002), Kirtikar and Basu (1984), Rastogi and Mehrotra (1991, 1995), Watt (1972).

Fagonia schweinfurthii Hadidi.

Syn.: *F. cretica* Auct.

Family: Zygophyllaceae

Part/s Used	:	Leaves.
Utilization	:	Decoction of fresh leaves said to cure cough and asthma.
Active Constituents	:	Alanine, arginine, glycine, isoleucine, leucine, lysine, phenylalanine, praline, tyrosine and valine isolated.
References	:	Rastogi and Mehrotra (1991, 1995), Singh and Pandey (1998).

Ferula assafoetida Linn.

Syn.: F. alliacea Boiss.; F. foetida Regel.

Family: Apiaceae

Sans. and H. Names	:	Sans.: Balhika, Hingu;
		H.: Hing.
Part/s Used	:	Roots and Gum.
Utilization	:	Stimulates respiratory tract; useful in asthma, chronic bronchitis and whooping cough.
Active Constituents	:	40-60 per cent resin, 25 per cent gum and 12-18 per cent volatile oil; a sesquiterpenoid coumarin–foetidin isolated form root.
References	:	Ambasta (1986), Bhattacharjee (2001), Duke et al. (2002), Rastogi and Mehrotra (1995), Sharma (2003).

Ferula narthex Boiss.

Syn.: Narthex asafoetida Falconer.

Family: Apiaceae

Eng., Sans. and H. Names	:	Eng.: Devils Dung, Persian fennel Giant;
		Sans.: Balhika, Bhutari, Hinguka, Madhura;
		H.: Hing, Hingra.
Distribution	:	Native of Kashmir in India.
Part/s Used	:	Gum and Resin.
Utilization	:	Asafoetida-a resinous and scented substance obtained from rhizome or roots, considered stimulant for respiratory system and very effective in pneumonia and bronchitis in children. Gum useful in asthma and whooping cough in spasmodic affection.
Active Constituents	:	Rhizome affords oleo gum 25 per cent and also volatile oil.
References	:	Bhattacharjee (1998), Jain (1968), Kirtikar and Basu (1984), Raju (2000), Sharma (2003), Watt (1972).

Ficus rumphii Blume

Syn.: C. cordifolia Roxb.; Urostiga rumphii Miq.; U. cordifolium Miq.

Family: Moraceae

H. Names	:	H.: Gagjaira, Gajiun, Gajna, Kabar, Pilkham, Pipul.

Distribution	:	N. India, Assam, C. India, W. Peninsula and S. India.
Part/s Used	:	Fruits.
Utilization	:	Extract of fruits considered useful to alleviate asthma.
References	:	Ambasta (1986), Kirtikar and Basu (1984), Watt (1972).

Ficus heterophylla Linn. f.

**Syn.: *F. scabrella* Roxb.; *F. truncata* Vahl.; *F. denticulata* Vahl.;
F. rufescens Vahl.; *F. truveata, repens, rufescens* Ham.;
F. aquatica Koenig.; *F. scabrella and heterophylla* Roxb.; *F. repens* Willd;
F. repens Roxb.; *F. rubifolia* Griff.**

Family: Moraceae

Sans. Name	:	*Sans.*: Trayamana.
Distribution	:	Throughout hotter parts of India.
Part/s Used	:	Root-bark.
Utilization	:	Pulverized root–bark mixed with coriander used for cough and asthma.
References	:	Ambasta (1986), Kirtikar and Basu (1984), Watt (1972).

Ficus nervosa Heyne *ex* Roth.

Family: Moraceae

Part/s Used	:	Fruits.
Utilization	:	10ml extract of fruits given twice daily for relief from asthma.
Reference	:	Prakash and Mehrotra (1999).

Foeniculum vulgare Mill.
[Pl. 3A]

Syn.: *F. officinale* All.; *F. capillaceum* Gillb.

Family: Apiaceae

Eng., Sans. and H. Names	:	*Eng.*: Fennel, Sweet fennel;
		Sans.: Bhuripushpa, Madhurica, Madhurika;
		H.: Baresaunf, Sont, Saunf.
Distribution	:	N. India. Maharashtra. Gujarat. Karnataka.
Part/s Used	:	Leaves and Seeds.
Utilization	:	Infusion of dry leaves recommended for cough. Seeds cure asthma.

Active Constituents	:	Seeds yield d- pinene, camphene d-L–phellandrene, dipentine, 50-60 per cent anethol fenchone, methyl chavicol, aldehyde and anisic acid.
Biological Activity	:	Antiasthmatic activity found to be positive.
References	:	Ambasta (1986), Bhattacharjee (1998), Jain *et al.* (1995), Nunez and Castro (1995), Sharma (2003).

Fritillaria imperialis Linn.
Family: Liliaceae

Eng. Names	:	*Eng.*: Crown Imperial.
Distribution	:	Kashmir–abundant near Baramulla at 1,700-3,000m.
Part/s Used	:	Corm.
Utilization	:	Considered useful in asthma.
Active Constituents	:	Main constituents are isobaimonidine, a new pimaradiene, a diterpne-oblongifolic acid, steroidal alkaloid–ebeinone, eduardine, edpetilidine, verticinone 7 isoverticine, cevarin, cevacine and imperialin.
References	:	Ambasta (1986), Rastogi and Mehrotra (1995).

Fritillaria roylei Hook.
Family: Liliaceae

Distribution	:	W. Temperate Himalaya 2,650–4,350m from Kashmir-Kumaon.
Part/s Used	:	Rhizome.
Utilization	:	Bulbs powdered and boiled with orange skin used for tuberculosis and asthma.
Active Constituents	:	Bulb contains peimine, peiminine, peimisine, peimiphine, peimitidine and natural propeimin and a stereol.
References	:	Chopra *et al.* (1956), Mehrotra *et al.* (1995), Sharma (2003).

Galeopsis tetrrahit Linn.
Family: Lamiaceae

Eng. Name	:	*Eng.*: Common Hemp nettle.
Distribution	:	Sikkim and Kashmir.
Part/s Used	:	Whole Plant.
Utilization	:	Infusion said to be useful in pulmonary troubles.

Active Constituents	:	Galiridoside isolated.
References	:	Ambasta (1986), Chopra *et al.* (1956), Rastogi and Mehrotra (1991).

Garuga pinnata Roxb.
Syn.: *G. madagascariensis* DC.
Family: Burseraceae

H. Names	:	*H.*: Ghogar, Kharpat, Kaikar.
Distribution	:	Widely distributed throughout India.
Part/s Used	:	Leaves.
Utilization	:	Leaf extract with honey or sugar given for asthma.
Active Constituents	:	Alcohol–6–propyltetradecan–7- ol and garuganin I isolated from leaves.
References	:	Ambasta (1986), Chhetri (1994), Kirtikar and Basu (1984), Lalramnghinglova (1996), Rastogi and Mehrotra (1990), Shah *et al.* (1981), Watt (1972).

Gentianella moorcroftiana Airy–Shaw
Syn.: *Gentiana moorcroftiana* Wall. *ex* G. Don
Family: Gentianaceae

Eng. Name	:	*Eng.*: Moorcroft's Gentian.
Distribution	:	W. Himalaya: 2,700-4,800m.
Part/s Used	:	Aerial Plant Parts.
Utilization	:	2–3g powdered aerial plant part, given with water, thrice a day, for cough.
References	:	Bennet (1987), Sood *et al.* (2001).

Geranium partense Linn.
Family: Gentianaceae

Eng. and H. Names	:	*Eng.*: Crane's bill, Crowfoot, Grace of God;
		H.: Banda.
Distribution	:	Temperate W. Himalaya, Europe and N. Asia.
Part/s Used	:	Whole Plant.
Utilization	:	Powdered plant (1\2 teaspoon) given with water twice daily to treat cough.
Active Constituents	:	Indotannin, isokempferid, hexahydroflavone isolated from plant.
References	:	Kirtikar and Basu (1984), Sood *et al.* (2001).

Ginkgo biloba Linn.

Family: Ginkgoaceae

Eng. Name	:	*Eng.*: Maidenhair Tree.
Distribution	:	Cultivated in Indian gardens; particularly on hills.
Part/s Used	:	Fruits.
Utilization	:	Fruits considered useful for bronchial complaints.
References	:	Bhattacharjee (1998), Sharma (2003).

Gleichenia dichotoma Willd.

Family: Gleicheniaceae

Distribution	:	Mountains of Southern India, Kumaon and Sikkim.
Part/s Used	:	Rhizome.
Utilization	:	Fronds given as a cure for asthma
Reference	:	Caius (2003).

Globba bulbifera Roxb.

Syn.: *G. marantina* L.

Family: Zingiberaceae

Part/s Used	:	Roots.
Utilization	:	1 tablet prepared by pulverizing 10g of its roots in combination with 10g of *Piper nigrum* given, twice daily for 15 days, for curing asthma and cough.
References	:	Bennet (1986), Tarafdar and Chaudhary (1981).

Globba marantina Linn.

Syn.: *G. bulbifera* Roxb.

Family: Zingiberaceae

Part/s Used	:	Rhizome.
Utilization	:	Extract of rhizome with a pinch of black pepper and 'ajowan' (*Trachyspermum ammi*) given, twice a day, for cough and asthma.
References	:	Brahmam and Saxena (1990), Singh and Pandey (1998), Tripathi and Goel (2001).

Gloriosa superba Linn.

Syn.: *G. angulata* Schum.; *Methonica superba* Lam.

Family: Liliaceae

Eng., Sans. and H. Names	:	*Eng.*: Climbing lily;

Sans.: Agnimukhi, Agnisikkha, Ananta, Dipta, Halmi, Languli, Sharapushpi;

H.: Kalihari, Kariari, Karihari, Kulhari, Languli.

Distribution	:	Throughout Tropical India.
Part/s Used	:	Leaves.
Utilization	:	Considered useful in asthma.
Active Constituents	:	Young leaves contain chelidonicacid, colchicines, dimethylcolchicine, N-formylde acetylcolchicine and lumicolchicine.
References	:	Chopra *et al.* (1980), Kirtikar and Basu (1984), Tarafdar and Chaudhary (1981), Watt (1972).

Glycyrrhiza glabra Linn.

Family: Fabaceae

Eng., Sans. and H. Names	:	*Eng.*: Licorice, Liquorice.
		Sans.: Madhuka, Jalayashti;
		H.: Jethimadh, Muthatti.
Distribution	:	Cultivated especially in Sub- tropical and tropical regions of India.
Part/s Used	:	Roots and Rhizome.
Utilization	:	Hot decoction of roots drunk for cough and bronchitis. Crude drug as well as its dried aqueous extract used in bronchial troubles along with *Viola pilosa*, *Adiantum lunulatum* and *Justicia adhatoda*. Rhizome chewed alone or with betel leaf relieves cough promptly.
Active Constituents	:	Principle constituents are glycyrrhizin, glycyrrhetol, glabrolide, glabrol, glyzarin, liquiritin, licoflavonol, licoisoflavones, licochalcones, herniarin and liqcomarin (rhizome), glycyrrhetic acid (1.84-2.75 per cent), glycyrrhizin (4-11 per cent) and total flavonoids (0.0282 -0.818 per cent), liquititigenin, isoliquiritigenin and 4-hydroxychalcone; glycyrrhisoflavone and glycyrrhisoflavone (roots).
References	:	Ambasta (1986), Gill and Nyawuame (1994), Jain (1968), Kirtikar and Basu (1984), Sharma and Sood (1997), Sharma (2003), Shome *et al.* (1996), Sinha and Sinha (2001), Rastogi and Mehrotra (1993, 1995).

Gymnema sylvestre R.Br.

Syn.: *Periploca sylvestris* Willd; *Asclepias geminata* Roxb.
Family: Asclepiadaceae

Eng., Sans. and H. Names	:	*Eng.*: Periploca of the Woods, Small Indian Ipe cacuanha;
		Sans.: Avartini, Chakshu, Karnika;
		H.: Gurmar, Merasing, Meshasingi.
Distribution	:	C. and S. India from Konkan to Travancore.
Part/s Used	:	Fruits.
Utilization	:	Fruit considered good in bronchitis.
Active Constituents	:	Gymnemagenin isolated from fruits.
References	:	Kirtikar and Basu (1984), Rastogi and Mehrotra (1993), Watt (1972).

Gynandropsis gynandra (Linn.) Briquet

Syn.: *G. pentaphylla* DC.; *Cleome gynandra* sp.
Family: Capparidaceae

Sans. and H. Names	:	*Sans.*: Surjavarta;
		H.: Gandhuli, Hulul.
Distribution	:	Common weed throughout warmer parts of India.
Part/s Used	:	Seeds.
Utilization	:	Infusion of seeds helps in relieving cough and bronchial affections.
Active Constituents	:	5, 7–Dihydrooxychromone, 5–hydroxyl -3, 7, 4'-trimethoxyflavone and luteolin isolated from seeds.
References	:	Das and Agarwal (1991), Kirtikar and Basu (1984), Rastogi and Mehrotra (1991).

Hedychium spicatum Ham. *ex* Sm.

Family: Zingiberaceae

Sans. and H. Names	:	*Sans.*: Amlaharidra, Durva, Gandha, Gandharika;
		H.: Gandhapalashi, Sitruti.
Distribution	:	Common in Panjab and Himalaya.
Part/s Used	:	Rhizome.
Utilization	:	Rootstock useful in asthma and bronchitis.

Active Constituents	:	A diterpene–6–oxo–labda–7, 11, 13–trine -16–oic acid lactone and hedychenone, 7 -hydroxyhedychenone isolated rom rhizome.
References	:	Kirtikar and Basu (1984), Malhotra and Balodi (1984), Rastogi and Mehrotra (1991, 1993, 1995), Sharma *et al.* (1979), Watt (1972).

Hedychium villosum Wall.

Family: Zingiberaceae

Part/s Used	:	Rhizome.
Utilization	:	Juice of crushed rhizome taken as effective remedy for asthma and cough.
Reference	:	Lalramnghinglova (1999).

Hedyotis umbellata (Linn.) Lam.

Syn.: *Oldenlandia umbellata* Linn.

Family: Rubiaceae

Eng. and H. Names	:	*Eng.*: Hay Root, Indian Madder;
		H.: Chirval.
Part/s Used	:	Leaves, Roots and Whole Plant.
Utilization	:	Pounded leaves and roots used as a remedy for asthma and bronchitis. Plant juice along with goat milk given to cure tuberculosis.
References	:	Ambasta (1986), Upadhyay and Chauhan (2000)

Helianthus annuus Linn.

Family: Asteraceae

Eng., Sans. and H. Names	:	*Eng.*: Common Sunflower, Marigold of Peru;
		Sans.: Suryamukhi, Suryavarta;
		H.: Harhuja, Surajmukhi.
Distribution	:	Cultivated in India.
Part/s Used	:	Flowers-heads, Leaves and Seeds.
Utilization	:	Boiled flower-heads used for pulmonary affections. A tincture of florets and leaves recommended in combination with balsamics for treating bronchiectasis. Infusion of roasted seeds admirable for whooping cough, bronchial and pulmonary affections, cough and cold.

Active Constituents	:	These are linoleic acid (seeds),echinocystic acid (petals), 3 saponins- helianthosides A,B and C kaurenic and trachylobanic acid, 2 Ψ- taraxene derivatives-heliantriol C and heliantriol F (flowers), niveusin C, agrophyllin and β–hydroxyl–3–dehydro–desoxy-fruticin (trichomes of leaves).
References	:	Ambasta (1986), Caius (2003), Das and Agarwal (1991), Kirtikar and Basu (1984), Rastogi and Mehrotra (1990, 1991, 1995), Singh *et al.* (1996).

Helichrysum melaleucum Rchb. *ex* Holl.
Family: Asteraceae

Part/s Used	:	Leaves and Shoots.
Utilization	:	Infusion of leaves and shoot recommended for bronchitis.
Reference	:	Nunez and Castro (1995).

Helicteres isora Linn.
Syn.: *H. chrysocalyx* Miq.; *H. roxburghii* G. Don
Family: Sterculiaceae

Eng. Name	:	*Eng.*: East Indian Screw Tree.
Distribution	:	Throughout C. and W. India.
Part/s Used	:	Roots and Fruits
Utilization	:	Considered useful for cough, cold and asthma.
Active Constituents	:	Cucurbitacin and isocucurbitacin identified in roots.
References	:	Asolkar *et al.* (1992), Mudgal and Pal (1980), Rastogi and Mehrotra (1993, 1995), Singh and Pandey (1980), Watt (1972).

Hemidesmus indicus (Linn.) Schultes
Syn.: *H. wallichii* Miq.; *Periploca indica* Willd; *Asclepias pseudosara* Roxb.
Family: Asclepiadaceae

Eng., Sans. and H. Names	:	*Eng.*: False Sarsaparilla, Indian Sarsaparilla;
		Sans.: Ananta, Bhadra, Sugandha;
		H.: Hindisalsa, Magrabu.
Distribution	:	Upper Gangetic plain, Eastwards to Bengal and Sundribans.
Part/s Used	:	Leaves.

Utilization	:	Hot infusion of dried leaves with sweet milk useful in chronic cough.
Active Constituents	:	Rutin isolated from leaves.
References	:	Islam (2000), Kirtikar and Basu (1984), Rastogi and Mehrotra (1991, 1995), Shah *et al.* (1981), Tarafdar (1986), Watt (1972).

Hepatica acutiloba DC.
Family: Ranunculaceae

Part/s Used	:	Leaves and Roots.
Utilization	:	Tea prepared from leaves considered useful for treating cough. Roots chewed to relieve persistent cough.
Reference	:	Bhattacharjee (1998).

Heracleum maximum Bartr.
Family: Apiaceae

Part/s Used	:	Roots.
Utilization	:	Roots prescribed in asthma.
Reference	:	Bhattacharjee (1998).

Hibiscus rosa-sinensis Linn.
[Pl. 3B]
Family: Malvaceae

Eng., Sans. and H. Names	:	*Eng.*: Chinese Hibiscus, Shoe flower.
		Sans.: Arkapriya, Aruna, Japa, Java,
		H.: Jasut, Jasum.
Distribution	:	Throughout India. Commonly planted as ornamental in house gardens.
Part/s Used	:	Flowers and Roots.
Utilization	:	Decoction of flowers lessens bronchial catarrh. Roots valuable in cough.
Active Constituents	:	2 cyclopeptide alkaloids I and II, quercetin-3-diglucoside, 3, 7- diglucoside, cyanidin-3, 5- diglucoside and cyaniding-3-sophoroside-5-glucoside isolated from flowers.
References	:	Ambasta (1986), Caius (2003), Gill *et al.* (1993), Kirtikar and Basu (1984), Rastogi and Mehrotra (1991, 1995), Sharma (2003).

Hippophae rhamnoides Linn.

Family: Elaeagnaceae

Eng. Names	:	*Eng.*: Common Sea Buckthorn, Sallow-Thorn.
Distribution	:	Drier ranges of N.W. Himalaya upto 2,135-4,572m.
Part/s Used	:	Fruits.
Utilization	:	Pulverized berries prescribed for tuberculosis. Syrup prepared from fruits useful in lung complaints.
Active Constituents	:	Ripe berries rich in humnin, carotene, ascorbic acid, dehydroascorbic acid, fatty oil (0.2 per cent), flavonoids– isorhamnetin and kaempferol, isoaharmnitol, isorhamnetin–3–β–D- glucoside, isoarhamnetin–3– rutinoside, isoarhamnetin–3–β- Dglucosido–α–L - rhamnoside, isorhamnetin–3–β–sophoroside -7–L– rhamnoside.
References	:	Ambasta (1986), Das and Agarwal (1991), Rastogi and Mehrotra (1990, 1991), Sharma and Sood (1997), Sood *et al.* (2001).

Hiptage benghalensis (Linn.) Kurz.

Syn.: *Banisteria benghalensis* L.; *H. madalota* Gaertn.; *H. parvifolia* W. and A.

Family: Malphighiaceae

Sans. and H. Names	:	*Sans.*: Atimukta, Bhushana, Kamuka, Madhvi;
		H.: Kampti, Madholota, Madmalti.
Distribution	:	Konkan. W. Ghats. North Kanara.Chennai. Mt. Abu. Kumaon. E. Bengal. Assam. Andaman.
Part/s Used	:	Leaves and Bark.
Utilization	:	Leaves and bark effective in cough and asthma.
Active Constituents	:	Friedelin, epifriedelinol, octacosanol, α- amyrin, β- sitosterol and its glucoside isolated from stem.
References	:	Bennet (1987), Kirtikar and Basu (1984), Rastogi and Mehrotra (1995).

Holarrhena antidysenterica Wall.

Syn.: *H. pubescens* Wall ex G.Don; *H. codaga* G.don; *H. malaccensia* Wight.; *Echites antidysenterica* Roxb.; *Wrightia antidysenterica* Grah.

Family: Apocynaceae

Eng., Sans. and H. Names	:	*Eng.*: Conessi tree, Kurchee tree, Tellicherry tree;

	Sans.: Indradu, Kalinga, Kutaja, Pandura;
	H.: Karchi, Karra, Kora, Kurchi.
Distribution	: Tropical Himalaya upto 1,000m and drier parts of India.
Part/s Used	: Root, Bark and Leaves.
Utilization	: 1-2 pills made from paste of stem–bark grounded with black pepper considered effective by Konda, Reddis, Koyas and Nuka Doras for cough and cold, twice a day, for 3 days. 2 tsp thick decoction of its roots boiled with equal quantity of fruits of *Terminalia chebula, T. bellirica* and *Solamum surattense* given, twice daily for a week, for cough and cold. Dried bark of plant constitutes the drug kurchi.
Active Constituents	: Bark contains conessine, kurchicne, kurchine, conamine, concurchine and holarrhenine.
Biological Activity	: Alkaloid conessine present in the bark has been found to retard the growth of *Tubercular bacilli.*
References	: Ambasta (1986), Amimuddin and Girach (1996), Rao and Henry (1996), Islam (2000), Jain (1968), Maheshwari *et al.* (1997), Pal (1999), Sharma (2003), Thakur *et al.* (1989), Varma (1997), Watt (1972).

Holoptelea integrifolia (Roxb.) Planch
Syn.: *Ulmus integrifolia* Roxb.
Family: Ulmaceae

Eng., Sans. and H. Names :	*Eng.*: Indian Elm;
	Sans.: Chirabilva;
	H.: Banchilla, Begana, Chilbil, Chilmil, Dhamina.
Distribution	: Outer ranges of Himalaya ascending to 700m, often planted in N. and C. India.
Part/s Used	: Stem-bark.
Utilization	: Extract of crushed stem-bark with garlic (*Allium sativum*) considered useful for asthma by Koyas and Valmikis–2 tsp, twice a day, till cure.
Active Constituents	: 2–aminonaphthaquinone (0.001), friedelin (0.002), epifriedetinol (0.005), β–sitosterol (0.004) and its β–D–glucoside (0.01 per cent), Friedelan–3–β–ol isolated from stem-bark.
References	: Henry and Rao (1996), Kirtikar and Basu (1984), Rastogi and Mehrotra (1991, 1995), Watt (1972).

Holostemma annularis K. Schum.

**Syn.: *H. rheedii* Wall; *H. rheediamum* auct. non Spreng.;
H. fragrans Wall.; *H. brunonianum* Royle; *H. adokodien* Roem and Sch.;
Asclepias annularis Roxb.; *Sarcostemma annulare* Roth.;
Gamphocarpus volubilis Herb.**

Family: Asclepiadaceae

Sans. and H. Names	:	*Sans.*: Arkapushpi, Jivanti, Kshirini;
		H.: Chhirvel.
Distribution	:	Tropical Himalaya and W. Peninsula.
Part/s Used	:	Roots.
Utilization	:	Decoction of roots employed by Santals for relief from cough and bronchitis.
Active Constituents	:	α–amyrin, lupeol, β–sitosterol, alanine, aspartic acid, glycine, serine, threonine and valine isolated from roots.
References	:	Ambasta (1986), Kirtikar and Basu (1984), Rastogi and Mehrotra (1993), Watt (1972).

Hordeum vulgare Linn.

Family: Poaceae

Eng., Sans. and H. Names	:	*Eng.*: Barely;
		Sans.: Akshata, Divya, Hayapriya, Kanchuki, Praveta, Shaktu, Sitashuka, Yava;
		H.: Jau, Jav, Jawa, Suj.
Distribution	:	Cultivated chiefly in N. India and upto 4,350m in the Himalaya.
Part/s Used	:	Grains.
Utilization	:	Decoction of barley grains mixed with honey employed for bronchial coughs.
Active Constituents	:	Starch, sugars, fats, proteins- albumin, globulin, prolamin, glutilin, orientoside, orientin, vitexin isolated from grains.
References	:	Anonymous (1999), Caius (2003), Kirtikar and Basu (1984), Rastogi and Mehrotra (1990).

Hornstedtia costata (Roxb.) Schum.

Syn.: *Amomum costatum* Benth.

Family: Zingiberaceae

Part/s Used	:	Seeds.

Utilization	:	Powdered seeds considered good for asthma and pulmonary diseases.
References	:	Ambasta (1986), Tripathi and Goel (2001).

Humboldtia unijuga Bedd.
Family: Fabaceae

Part/s Used	:	Roots.
Utilization	:	Root paste good for asthma.
Reference	:	Radhakrishnan *et al.* (1996).

Hymnostegia afzeli (Oliv.) Harms
Family: Fabaceae

Part/s Used	:	Bark.
Utilization	:	Powdered astringent bark prescribed for cough.
Active Constituents	:	Tannins and alkaloids.
Reference	:	Gill and Nyawuame (1994).

Hypericum japonicum Thunb. ex Murr.
Syn.: *H. japonicum* var. *calyculatum* R. Keller; *H. japonicum* var. *simplicius* R. Keller; *H. japonicum* var. *majus* Fyson
Family: Papaveraceae

Eng. and H. Names	:	*Eng.*: Amber, Cammock, Grace of god;
		H.: Basant, Dendhu.
Distribution	:	Temperate W. Himalaya: 2,000-3,000m.
Part/s Used	:	Whole Plant.
Utilization	:	Infusion of the plant given beneficially for chronic catarrh of lungs. Plant extract anti–asthmatic.
Active Constituents	:	Plant affords saropeptate, sarothralen C, sarothralen D, a flavone glycoside-3, 5, 7, 3', 4'- pentahydroxy–flavone–7–O–rhamnoside and japonicine A.
Biological Activity	:	Aqueous extract of plant found to inhibit growth of *Mycobacterium tuberculosis*.
References	:	Ambasta (1986), Bennet (1987), Chopra *et al.* (1980), Kirtikar and Basu (1984), Rastogi and Mehrotra (1993, 1995).

Hyocyamus niger Linn.

Family: Solanaceae

Eng., Sans. and H. Names	:	*Eng.*: Black Henbane, Henbane, Stinking Roger;
		Sans.: Henbane, Parasikaya, Shyama;
		H.: Ajvanyan, Khurasani.
Distribution	:	W. Himalaya from Kashmir to Kumaon 1,500- 3,000m.
Part/s Used	:	Leaves and Fruits.
Utilization	:	Decoction made from leaves and fruit admirable for whooping cough and asthma.
Active Constituents	:	Hyosyamine and hyoscine; scopolamine with little atropine, hyoscypikrin, skimmianine, apohyoscine, tropine, α- and β- belladonine, cuscohygrine, 6β–hydroxyhyoscyamine, apoatropine, hyoscyamine–N–oxide isolated from roots, leaves and stem.
References	:	Ambasta (1986), Bhattacharjee (2001), Chandrasekhar and Srivastava (2003), Jain (1968), Kirtikar and Basu (1984), Rajan *et al.* (1999), Rastogi and Mehrotra (1991), Sharma (2003), Sharma and Sood (1997), Thakur *et al.* (1989).

Hyssopus officinalis Linn.

Family: Lamiaceae

Eng. and H. Names	:	*Eng.*: Hyssop;
		H.: Zufahyabis.
Distribution	:	Himalaya (Kashmir to Kumaon): 2,600- 3,700m.
Part/s Used	:	Flowers and Whole Plant.
Utilization	:	Infusion of the herb given, 2 or 3 times a day, for pulmonary catarrh and asthma. Preparation from flowers considered useful for asthma, cough, cold and lungs complaints. Hissop's oil extracted from flowers in one or 2 drops promotes expectoration in bronchial catarrh and asthma.
Active Constituents	:	These are volatile oil, codine, tannins, carotene, xanthophylls, fat, sugar and ursolic acid (herb), betone-1-pinocamphene, α-pinene, L-pinene, camphene, 1-pinocamphenol; myrtenol methyl ether, methyl myrtenate, cis- pinic acid, (+)2-hydroxyisopinocamphene, cis- pinonic acid and β-pinene (oil).
References	:	Bhattacharjee (2000), Kirtikar and Basu (1984), Rastogi and Mehrotra (1991), Sharma (2003), Sharma and Sood (1997), Sinha and Sinha (2001), Watt (1972).

Iberis amara Linn.
[Pl. 3C]
Syn.: *Iberis coronaria* Hort.
Family: Brassicaceae

Eng. Names	:	*Eng.*: Common Annual Candytuft.
Distribution	:	Cultivated as garden ornamental during winter.
Part/s Used	:	Seed.
Utilization	:	Seeds considered useful for asthma and bronchitis.
References	:	Ambasta (1986), Chopra *et al.* (1980), Rajan *et al.* (1999).

Indigofera hirsuta Linn.
Family: Fabaceae

Eng. Names	:	*Eng.*: Hairy indigo.
Part/s Used	:	Whole Plant.
Utilization	:	Decoction of plant or its powder given for whooping cough and bronchitis.
References	:	Ambasta (1986), Chopra *et al.* (1980), Gill and Nyawuame (1994).

Indigofera pulchella Roxb.
Family: Fabaceae

H. Names	:	*H.*: Hakna, Sakena.
Distribution	:	Throughout Himalayan tract. Hills of India ascending to 1,700m.
Part/s Used	:	Flowers and Roots.
Utilization	:	Extract of roots crushed with black peppers considered beneficial for asthma by Jatapus and Khonds, 2 tsp twice a day for 5 days. Decoction of flowers used for cough by Santals.
References	:	Ambasta (1986), Chopra *et al.* (1980), Henry and Rao (1996), Kirtikar and Basu (1984), Watt (1972).

Indigofera tinctoria Linn.
Family: Fabaceae

Eng., Sans. and H. Names	:	*Eng.*: Common indigo;
		Sans.: Nili, Nilika, Nilla;
		H.: Nil.

Distribution	:	Widely cultivated in many parts of country.
Part/s Used	:	Roots and Whole Plant.
Utilization	:	Extract of plant given for bronchitis. 1 tsp decoction of roots (50g) crushed with black pepper (5g) taken, twice a day for 21 days, for asthma.
References	:	Ambasta (1986), Anonymous (1999), Deshmukh *et al.* (1999), Kirtikar and Basu (1984), Rastogi and Mehrotra (1995), Rao and Pullaiah (2001).

Inula racemosa Hook.f.

Family: Asteraceae

Distribution	:	Cultivated areas, forest clearings, shrubberies in N. W. Himalaya from Kashmir to Uttaranchal: 1,600-4,200m.
Part/s Used	:	Roots.
Utilization	:	Root extract antiasthmatic.
References	:	Sharma (2003), Rastogi and Mehrotra (1990, 1991, 1993), Watt (1972).

Jasminum angustifolium Vahl.

Family: Oleaceae

Eng., Sans. and H. Names	:	*Eng.*: Wild jasmine;
		Sans.: Asphota, Vanamalli;
		H.: Banmallica.
Distribution	:	Deccan Peninsula and S. Travancore.
Part/s Used	:	Leaves.
Utilization	:	Preparation of leaves effective for controlling bronchitis and asthma.
References	:	Ambasta (1986), Bhattacharjee (2001), Kirtikar and Basu (1984), Sharma (2003), Watt (1972).

Jasminum arborescens Roxb.

Syn.: *J.roxburghianum* Wall.; *J. latifolium* Roxb.

Family: Oleaceae

Eng., Sans. and H. Names	:	*Eng.*: Tree jasmine;
		Sans.: Saptala;
		H.: Bela chameli.
Distribution	:	Tropical N.W. Himalaya and Deccan Peninsula: 3,000-3,000m.

Part/s Used	:	Leaves.
Utilization	:	Extract of leaves enters into preparation of emetics used in bronchial obstructions.
References	:	Ambasta (1986), Sharma (2003).

Jatropha curcas Linn.

Family: Euphorbiaceae

Part/s Used	:	Stem–bark.
Utilization	:	Stem-bark powdered along with a pinch of salt considered good for cough by Konda Reddis and Koya (2-3 pills, thrice a day for 3 days).
Reference	:	Rao and Henry (1996).

Juniperus communis Linn.

Family: Cupressaceae

Eng., Sans. and H. Names	:	*Eng.*: Common juniper;
		Sans.: Vapusha;
		H.: Aaraar, Abhal.
Distribution	:	Himalayas from Kumaon westwards between 1,500-4,200m.
Part/s Used	:	Fruits.
Utilization	:	Extract of fruits given for cure of pulmonary catarrh and asthma.
Active Constituents	:	Fruits yield volatile oil, sugar, and resin (10 per cent) juniperin, fixed oil, protein, wax, gum, pectins, organic acids and salts
References	:	Ambasta (1986), Bhattacharjee (1998), Singh (1996).

Juniperus macropoda Boiss.

Family: Cupressaceae

Eng. Names	:	*Eng.*: Himlayan Pencil Cedar, Indian Juniper.
Distribution	:	Dry rivers valleys and Uttarakhand: 2,400–4,300m.
Part/s Used	:	Fruits.
Utilization	:	1 tsp decoction of fruits prescribed for asthma, at bed time, for 10-15 days.
Active Constituents	:	Fruits yield essential oil constituting sugiol, junipodin and glycoside junipin.
References	:	Ambasta (1986), Kaul (1997), Singh (2003).

Juniperus virginiana Linn.

Family: Cupressaceae

Eng. Names	:	*Eng.*: Pencil cedar, Red cedar.
Part/s Used	:	Leaves, Twigs and Fruits.
Utilization	:	Inhalation of vapours from a hot decoction of leaves, twigs and fruits effective for bronchitis.
References	:	Ambasta (1986), Bhattacharjee (1998).

Justicia procumbens Linn.

Family: Acanthaceae

Utilization	:	Infusion of herb considered beneficial for asthma and cough.
Reference	:	Ambasta (1986).

Kaempferia galanga Linn.

Syn.: *Alpinia sessilis* Kon.
Family: Zingiberaceae

Eng., Sans. and H. Names	:	*Eng.*: Black thorn;
		Sans.: Chandramulika;
		H.: Chandramulo.
Distribution	:	Throughout the hotter parts of India.
Part/s Used	:	Rhizome.
Utilization	:	Powdered rhizome given with honey to cure cough and asthma, 2 times a day for a week.
Active Constituents	:	Chief constituents of rhizome are essential oil, ethyl p–methoxy–trans–cinnamate, deoxypodophyllotoxin, ethyl p–methoxy–trans–cinnamate and monoterpene ketone–cae–3–en -5–one.
References	:	Ambasta (1986), Islam (1996), Kirtikar and Basu (1984), Rastogi and Mehrotra (1991, 1995), Tripathi and Goel (2001), Watt (1972), Yusuf *et al.* (2002).

Kalanchoe laciniata (Linn.) Pers.

Family: Crassulaceae

Sans. Name	:	*Sans.*: Hemasagara.
Distribution	:	Bengal and Deccan.
Part/s Used	:	Leaves.

Utilization	:	Poultice of leaves considered useful for cough and cold.
Active Constituents	:	Active principles of leaves are bufadiennolide–bryophyllin B, bryophullol, bryophollone, bryophollenone, bryophynol and 2 homologous phenanthrene derivatives.
References	:	Jain and Tarafdar (1970), Kirtikar and Basu (1984), Kumar (2002), Rastogi and Mehrotra (1995).

Kalanchoe pinnata (Lamk.) Pers.

Syn.: *Bryophyllum calycinum* Salisb.; *B. pinnatum* Kurz.;
Cotyledon pinnatum Lamk.; *B. pinnatum* (Lam.) Oken.

Family: Crassulaceae

Sans. and H. Names	:	*Sans.*: Abtibhaksha, Parnabija;
		H.: Zakhmhaiyat.
Distribution	:	Throughout the hot and moist parts of India, particularly common in Bengal.
Part/s Used	:	Leaves.
Utilization	:	A single raw leaf eaten for 7 days for cough.
Active Constituents	:	These are p-coumaric, ferulic, syringic, caffeic and p-hydroxybenzoic acids, quercetins, kaempferol, malic, isocitric and citric acids.
References	:	Bennet (1987), Kirtikar and Basu (1984), Rastogi and Mehrotra (1993), Upadhye *et al.* (1994).

Kedrostis rostrata (Rottl.) Cogn.

Syn.: *Rhynchocarpa foetida* C.B. Clarke non Schrad.;
K. foetidissima Jacq. Trichosanthes foetidissima Jacq.

Family: Cucurbitaceae

Distribution	:	Gujarat, Konkan and Deccan.
Part/s Used	:	Roots.
Utilization	:	Powdered roots admirable for humoral asthma.
References	:	Ambasta (1986), Bennet (1987), Kirtikar and Basu (1984).

Kyllinga nemoralis (J.R. and G. Forst)
Danty *ex* Hutch and Dalziel

Family: Cyperaceae

Part/s Used	:	Whole Plant.

Utilization	:	Extract of plant, (2 tsp) given, thrice a day, for cough and cold.
Reference	:	Kshirsagar and Singh (2000).

Lactuca sativa Linn.

**Syn.: *L. scariola* Linn. var. *sativa* C.B. Clarke;
L. bracteata and *sativa* Wall.**

Family: Asteraceae

Eng. and H. Names	:	*Eng.*: Garden lettuce, Prickly lettuce;
		H.: Kahu, Khas, Salad.
Distribution	:	W. Himalaya: 3,000-4,000m.
Part/s Used	:	Seeds and Whole Plant.
Utilization	:	Decoction of fresh plant useful in cough, bronchitis and asthma; and that of seeds for chronic bronchitis.
Active Constituents	:	Active principles are campesterol, stigmasterol, siotosterol, 5–dehydroavenasterol, stigast -7–en–3β–ol and 7–dehydro–avenasterol (seeds), sesquiterpene lactones–3β- hydroxyl -11β,13 -dihydroacantho-spermolide and 3β,14-dihydroxy–11β,13-dihydroco-stunolide (aerial parts).
References	:	Ambasta (1986), Caius (2003), Kirtikar and Basu (1984), Rastogi and Mehrotra (1991, 1995), Rajan *et al.* (1999), Watt (1972).

Lactuca virosa Linn.

Family: Asteraceae

Eng. Name	:	*Eng.*: Bitter Lettuce.
Part/s Used	:	Latex.
Utilization	:	Considered useful in cough and asthma.
Active Constituents	:	Lactucin, lactucic acid, lactucopicrin, caoutchouc, albumen, mannite and certain inorganic substances isolated from latex.
References	:	Ambasta (1986), Greenish (1999), Watt (1972).

Laggera alata (Sch.) Bip.

Family: Asteraceae

Distribution	:	Tropical Himalaya.
Part/s Used	:	Leaves.

Utilization	:	Decoction of dry leaves effective in tuberculosis.
References	:	Jamir (1991), Sharma (2003).

Lannea coromandelica (Houtt.) Merr.

Syn.: *Calenum grande* O. Ktze.; *Lannea woodier* Parker; *Dialium coromandelicum* Houtt.; *Odina wodier* Roxb.

Family: Anacardiaceae

Sans. and H. Names	:	*Sans.*: Ajasringi, Jivala, Manjari;
		H.: Jhingan, Jingan, Kamlai, Moyen.
Distribution	:	Deciduous forests throughout India. Sub–Himalayan tract ascending to1, 350m.
Part/s Used	:	Bark.
Utilization	:	1 cup of decoction of bark taken, once daily for 3 days, for cough.
Active Constituents	:	Phlobatannins and unidentified hydroxyanthra-quinones, dl-epicatechin and leucocyanidin isolated from bark.
References	:	Kirtikar and Basu (1984), Rastogi and Mehrotra (1990), Sharma (2003), Upadhye *et al.* (1994).

Lantana camara Linn.

Family: Verbenaceae

Distribution	:	Naturalized in many part of India.
Part/s Used	:	Leaves.
Utilization	:	Leaf extract effective for tuberculosis.
Active Constituents	:	Lantoic acid,lantanilic acid, lnatadene D, camaroside, oleanolic acid, lantadene A, lantadene B, and icterogenin isolated from leaves.
Biological Activity	:	Alkaloidal fraction of leaves accelerated deep respiration.
References	:	Chopra *et al.* (1956), Rastogi and Mehrotra (1991, 1995), Seetharam *et al.* (1999).

Leea aequata Linn.

Syn.:*L. hirta* Roxb. *ex* Hornem; *L. scabra* Steud.; *L. Kurzii* Cl.

Family: Leeaceae

Sans. and H. Names	:	*Sans.*: Dasi, Kakajangha, Kakakala, Kakanasa;
		H.: Kakajangha, Kakjangha.

Distribution	:	Sikkim Himalaya.Assam. E. Bengal. Andamans.
Part/s Used	:	Roots.
Utilization	:	Roots considered useful in Bronchitis.
Biological Activity	:	Essential oil inhibits growth of *Mycobacterium tuberculosis* in concentration of 10 µg/cc,100µb/cc and 50 µg/cc respiration.
References	:	Ambasta (1986), Bennet (1987), Chopra *et al.* (1980), Kirtikar and Basu (1984), Watt (1972).

Leea compactiflora Kurz.

Syn.:*Leea robusta* Roxb.; *Leea aspera* Wall.; *L. trifoliata* Lawson; *L. bracteata* Cl.; *L. robusta* auct non Roxb.

Family: Leeaceae

Distribution	:	Sikkim Himalaya. Khasia Mountains: 350-1,700m.
Part/s Used	:	Roots.
Utilization	:	Dried tuberous roots used for asthma and tuberculosis; 2 tsp, twice a day with honey for 1 month.
References	:	Bennet (1987), Sinha and Dixit (2003), Watt (1972).

Lens esculenta Moench

Syn.: *Ervum lens* Linn.; *Cicer lens* Willd; *L. culinaris* Medik.

Family: Fabaceae

Sans. and H. Names	:	*Sans.*: Gabholika, Gurubija, Masuraka, Masurika;
		H.: Masur.
Distribution	:	A cold weather crop throughout India.
Part/s Used	:	Seeds.
Utilization	:	Seeds considered useful in bronchitis.
Active Constituents	:	Imidazole, p-coumaric acid, ferulic acid, 24 methylene-25-methylcholesterol, ciceritol, tricetin, luteolin, diglycosyldelphinidin and 2 proanthocyanidins isolated from seeds.
References	:	Kirtikar and Basu (1984), Rastogi and Mehrotra (1991, 1995), Watt (1972).

Leonurus cardiaca Linn.

Family: Lamiaceae

Distribution	:	Kumaon, Kashmir, Punjab and Kurrum valley.
Part/s Used	:	Aerial Plant Parts.

Utilization	:	Infusion considered useful for treating asthma.
Active Constituents	:	Principle constituents of aerial parts are ajugol, ajugoside, galirodoside, leonuroside, epigenin, kaempferol, quercetin, stachydrine, betonicine, luricine, leonurine, tannins, small amount of volatile oil and leocardin.
References	:	Bhattacharjee (2001), Chopra *et al.* (1956), Rastogi and Mehrotra (1990, 1993, 1995).

Lepidium sativum Linn.

Family: Brassicaceae

Eng., Sans. and H. Names	:	*Eng.*: Garden Crees;
		Sans.: Ashalika, Chandrika, Karavi, Nandini;
		H.: Chumsar, Halim, Hurf.
Distribution	:	Cultivated throughout India.
Part/s Used	:	Seeds and Whole Plant.
Utilization	:	Plant employed for asthma and cough. Seeds useful in bronchitis.
Active Constituents	:	Lepidine, sinapic acid ethyl ester, N'N- dibenzyl-thiourea and N'N–dibenzylurea, lepidimoide isolated from seeds.
References	:	Kirtikar and Basu (1984), Rastogi and Mehrotra (1990, 1995).

Leptadenia reticulata (Retz.) Wight and Arn.

**Syn.: *Cynanchum reticulatum* Retz.; *L. appendiculata* Dcne.;
L. imberbe Wight; *L. brevipes* Wight.; *C. asthmaticum* Ham.;
C. ovatum Thunb.; *Secamone canescens* Sm.**

Family: Asclepiadaceae

Sans. and H. Names	:	*Sans.*: Arkapushpi, Svarnajivanti;
		H.: Dori.
Distribution	:	Punjab and W. Peninsula.
Part/s Used	:	Leaves, Fruits and Latex.
Utilization	:	Latex inhaled for cold and leaves and fruits for asthma.
Active Constituents	:	Leptaculatin isolated from leaves.
References	:	Bhatt and Mitaliya (1999), Henry and Rao (1996), Kirtikar and Basu (1984), Rastogi and Mehrotra (1995), Watt (1972).

Leucas aspera Spreng.

Syn.: ***Leucas plukenetii* (Roth.) Spreng.; *Phlomis plukenetii* Roth.**

Family: Lamiaceae

H. Name	:	*H.*: Chota lalkhusa.
Distribution	:	More or less throughout India in the plains.
Part/s Used	:	Leaves, Flower and Whole Plant.
Utilization	:	Extract of leaves, (2 tsp, thrice a day) taken for cough and cold. Flowers mixed with honey prescribed for cough and cold in children. Decoction (20-40 ml) of whole plant mixed with a pinch of black pepper used in cough, cold and bronchitis.
References	:	Ambasta (1986), Chopra *et al.* (1956), Jha and Varma (1996), Kumar (2002), Maheshwari *et al.* (1997), Maya *et al.* (2003), Punjani (2002), Trivedi (2002).

Leucas cephalotes Spreng.

Syn.: ***Leucas capitata* Desf.; *Phlomis cephalotes* Roth.**

Family: Lamiaceae

Sans. and H. Names	:	*Sans.*: Dronapushpi, Kurumba, Palindi, Supushpi;
		H.: Deldona, Dhurpi Sag, Goma, Motapati.
Distribution	:	Himalaya: 600- 1,800m. Wastelands throughout the country.
Part/s Used	:	Leaves and Inflorescence.
Utilization	:	Syrup prepared from flowers, leaves and inflorescence after adding some sugar and water taken orally, twice daily, for 4–5 days for cough and cold.
Active Constituents	:	These are glycoside, β-sitosterol, and flavonoid, an essential oil, α-sitosterol and a glycoside (aerial plant parts).
References	:	Ambasta (1986), Anonymous (1999), Bhatt and Mitaliya (1999), Caius (2003), Kirtikar and Basu (1984), Sharma (2003), Singh and Pandey (1998), Watt (1972).

Leucas indica (L.) R.Br. *ex* Vatke.

**Syn.:*Leomurus indicus* L.; *Leucas lavandulifolia* Smith;
Leucas linifolia (Roth.) Spreng.; *Phlomis linifolia* Roth;
P. zeylanica Roxb.; *Leomurus indicus* Burm. var. *decipiens* Sm.**

Family: Lamiaceae

Distribution	:	Assam, Bengal, Deccan and W. coast from Konkan to Travancore.

Part/s Used	:	Whole Plant.
Utilization	:	A glassful of soup obtained by boiling plant body in water along with a little common salt taken, twice a day on empty stomach for cough and cold.
Active Constituents	:	Acacetin and chrysoeriol isolated from aerial parts.
References	:	Bennet (1987), Bhattacharjee (2001), Rastogi and Mehrotra (1995), Varma (1997), Watt (1972).

Leucas mollissima Wall. *ex* Benth. var. *scaberula* Hook.f.

Syn.: *Leucas pilosa* var. *pubescens* Benth.

Family: Lamiaceae

Distribution	:	Sub–tropical Himalaya, Khasia Hills, Konkan and C. India.
Part/s Used	:	Leaves and Twigs.
Utilization	:	Leaves and young twigs consumed as pot herb to cure cough.
References	:	Brahmam and Saxena (1990), Watt (1972).

Leucas unicaefolia R. Br.

Family: Lamiaceae

Distribution	:	Bengal. Punjab. Gujarat. Deccan.
Part/s Used	:	Whole Plant.
Utilization	:	Plant decoction considered useful for asthma and cough.
Active Constituents	:	Kaempferol and quercetin are major flavanoids isolated from leaves.
References	:	Bhatt *et al.* (1999), Kirtikar and Basu (1984), Rastogi and Mehrotra (1995).

Lindenbergia indica (Linn.) Kuntze.

Syn.: *L. urticaefolia* Lehm.; *L. ruderalis* (Retz.) Voigt.; *Dodartia indicia* L.; *L. polyantha* Royle *ex* Benth.; *L. muraria* (Roxb.*ex* D.Don) Bruehl.; *Stemodia muraria* Roxb. *ex*. D.Don

Family: Scrophulariaceae

Distribution	:	Throughout India ascending to 1,500m in hills.
Part/s Used	:	Whole Plant.
Utilization	:	Extract of plant useful for chronic bronchitis.

Active Constituents	:	2 steryl glycosides-α-L- rhamnopyranosyl $(1\rightarrow 4)$- β-D-glucopyranosyl $(1\rightarrow 3)$ sitosterol and furanosyl $(1\rightarrow 3)$ sitosterol, friedelan-3β-ol, oleanolic acid, 7-hydroxyflavone, quercetin, hydrocarbons (C 27-33), mannitol, apigenin, β-sitosterol, its glycoside and palmitate isolated from plant.
References	:	Ambasta (1986), Kirtikar and Basu (1984), Kumar (2002), Rastogi and Mehrotra (1993, 1995).

Lindera neesiana Benth.

**Syn.:*Aperula neesiana* Blume; *Benzoin neesianum* Nees;
Tetranthera neesiana Wall; *T. prunifolia* Wall.; *Laurus macrophylla* Don.**

Family: Lauraceae

Distribution	:	Temperate Himalaya: 2000-2,700m.
Part/s Used	:	Seeds.
Utilization	:	A small quantity of seed oil prescribed for cough and throat trouble.
Active Constituents	:	Active alkaloids isolated from seeds are methyl laurate (75.2 per cent), methyl caprate (13.3 per cent), methyl oleate (5.4 per cent), methyl myristate (2.4 per cent) and methyl palmitate (0.5 per cent).
References	:	Jain and Saklani (1992), Rastogi and Mehrotra (1995), Watt (1972).

Linum usitatissimum Linn.

Syn.: *L. trinervium* Roth.

Family: Linaceae

Eng., Sans. and H. Names	:	*Eng.*: Common flax, Flax, Flix, Linseed;
		Sans.: Devi, Hainwati, Malina, Uma- Chanaka;
		H.: Alsi, Tisi.
Distribution	:	Cultivated throughout India upto 2,000m.
Part/s Used	:	Seeds.
Utilization	:	A wine glass of linseed tea taken for cough and cold. Paste of boiled seeds and those of *Trigonella* mixed with honey prescribed for pneumonia.
Active Constituents	:	Seeds contain HCN glucoside–linamarin, 30-40 per cent of fixed oil, oleic, linoleic and linolenic acids; isofucosterol, cholesterol (2 per cent), campesterol (26

per cent), stigmasterol (7 per cent), sitosterol (41 per cent), 5-dihydro-avenasterol (13 per cent), cycloartenol (9 per cent).

References	:	Caius (2003), Das and Agarwal (1991), Greenish (1999), Kirtikar and Basu (1984), Rajan *et al.* (1999), Rastogi and Mehrotra (1991, 1995), Upadhyay and Chauhan (2002).

Lippia nodiflora (L.) Rich.

Syn.: *Verbena nodiflora* Linn.; *V. capitata* Forsk.; *Zapania nodiflora* Lamk.; *Lantana sarmentosa* Spreng.; *Phyla chinensis* Lour.

Family: Verbenaceae

Sans. and H. Names	:	*Sans.*: Agnijvala, Chitrapatri, Langali;
		H.: Jalapapili, Ludra.
Distribution	:	Throughout India.
Part/s Used	:	Whole Plant.
Utilization	:	Squeezed plant inhaled for treating cough and cold.
Active Constituents	:	Nodiflorin A, nodiflorin B, lactose, maltose, glucose, fructose and xylose and nodifloretin characterized from the plant.
References	:	Kirtikar and Basu (1984), Manandhar (1990), Rastogi and Mehrotra (1990), Watt (1972).

Litsea khasyana Meissn.

Family: Lauraceae

Part/s Used	:	Fruits.
Utilization	:	Powdered dry fruits along with a few fruits of *P. nigrum* L. and little sugar candy given for chronic bronchitis.
Reference	:	Chhetri (1994).

Lobelia erinus Linn.

Family: Campanulaceae

Distribution	:	Growing in NBRI garden.
Part/s Used	:	Fresh flowering plant.
Utilization	:	Recommended for asthma and cough.
Reference	:	Rajan *et al.* (1999).

Lobelia inflata Linn.

Family: Campanulaceae

Eng. Name	:	*Eng.*: Indian tobacco.
Distribution	:	Cultivated in India.
Part/s Used	:	Dried leaves.
Utilization	:	Leaves smoked to obtain relief from respiratory difficulties.
Active Constituents	:	Leaves contain lobeline, lobelinine, lobelanine, lobelanidine and isolobinine.
Biological Activity	:	Lobelinine acts as a respiratory stimulant; relaxes the bronchial muscles and thus dilates the bronchioles.
References	:	Ambasta (1986), Rajan *et al.* (2002), Bhattacharjee (1998), Greenish (1999).

Lobelia nicotianaefolia Hyne. *ex* Roth.

Syn.: *Rapuntium nicotianaefolium* Presl.; *L. pyramidalis* Wall.; *L. rosea* Wall.; *L. erecta* Hook. f. and Thoms.; *L. beddomeana* Wimmer; *L. pyramidalis* Wall; *L. Seguinii* var. *doniana* Wimmer.

Family: Campanulaceae

Eng., Sans. and H. Names	:	*Eng.*: Wild tobacco;
		Sans.: Devanla;
		H.: Narasal.
Distribution	:	Hilly regions of the Deccan Peninsula.
Part/s Used	:	Aerial parts of the plant.
Utilization	:	Shade dried aerial parts useful for asthma. Inhalation of vapours from hot decoction of stem effective in causing general relaxation.
Active Constituents	:	Lobeline (20 per cent) isolated from aerial plant parts.
References	:	Bennet (1987), Bhattacharjee (1998), Jain (1968), Sharma (2003), Prakash and Mehrotra (1987), Upadhye *et al.* (1994).

Lonicera spinosa (Jacquem. *ex* Decne) Walp.

Family: Caprifoliaceae

Part/s Used	:	Flowers.
Utilization	:	Decoction of flowers employed in treating asthma.
References	:	Singh *et al.* (1996).

Luffa acutangula (L.) Roxb.

Syn.: *Cucumis acutangulus* L.; *L. acutangula* var. *amara* (Roxb.) Cl.; *L. kleinii* W.and A.; *L. umbellata* (Klein *ex* Willd.) Roem.

Family: Cucurbitacceae

Eng., Sans. and H. Names	:	*Eng.*: Ridged or Ribbed Gourd;.
		Sans.: Dharaphala, Jalini, Jhongaka, Torai;
		H.: Jinga, Kali tori, Turi.
Distribution	:	Cultivated throughout the greater part of India.
Part/s Used	:	Leaves and Fruits.
Utilization	:	Dried powdered leaves and fruits administered for asthma.
Active Constituents	:	Cucurbitacins B and E and oleanolic acid isolated from fruits.
References	:	Ambasta (1986), Bennet (1987), Das and Agarwal (1991), Duke *et al.* (2002), Kirtikar and Basu (1984), Rastogi and Mehrotra (1993).

Luffa echinata Roxb.

Syn.:*L. bindaal* Roxb.; *Luffa echinata var. longistyla* Cl.

Family: Cucurbitaceae

Sans. and H. Names	:	*Sans.*: Deodalika, Devadali, Ghosha, Kadamba;
		H.: Bidali, Ghusarana, Vandala.
Distribution	:	Gujarat, Sind, Bihar, Dehradun, Bundelkhand and Bengal.
Part/s Used	:	Roots and Fruits.
Utilization	:	Roots and fruits considered good for bronchitis.
Active Constituents	:	Cucurbitacins B, Cucurbitacin E and a bitter saponin composed of triterpene acid, glucose, arabinose and rhamnose, datiscacin and cucurbitacin S-2-O-β-D-glucose isolated from fruits.
References	:	Ambasta (1986), Das and Agarwal (1991), Kirtikar and Basu (1984), Rastogi and Mehrotra (1990, 1995).

Lycopodium cernuum Linn.

Syn.: *Palhinhaea cernuua* (L.) Francoand

Family: Lycopodiaceae

Distribution	:	Throughout hilly regions of India and in S. Andaman Island.

Part/s Used	:	Whole Plant.
Utilization	:	Decoction of plant lessens cough and chest complaints.
Active Constituents	:	Cernuine, lycocernuine and a small quantity of nicotine isolated from plant.
References	:	Rastogi and Mehrotra (1990), Singh (1999).

Machilus macrantha Nees
Syn.: *Persea macrantha* (Nees) Kostermans.
Family: Lauraceae

Eng. Name	:	*Eng.*: Machilus.
Distribution	:	W. Peninsula.
Part/s Used	:	Bark.
Utilization	:	Decoction of bark given in asthma.
Active Constituents	:	An arabinoxylan composed of arabinose (73 per cent) and xylose isolated from bark.
References	:	Ambasta (1986), Rastogi and Mehrotra (1993).

Madhuca indica Gel.
Syn.: *Bassia latifolia* Roxb.; *Madhuca latifolia* Roxb.; *M. longifolia* (Koen.) McBride var. *latifolia* (Roxb.) Chev.; *B. longifolia* Koenig.
Family: Sapotaceae

Eng., Sans. and H. Names	:	*Eng.*: Butter tree, Mahua, S. Indian Mahua;
		Sans.: Madhuka;
		H.: Mahua, Mohwa.
Distribution	:	Plains and lower hills of India upto 1,200m.
Part/s Used	:	Flowers.
Utilization	:	Decoction of flowers given in cough and bronchitis. Also, pickle (2tsp) prepared from flowers given, twice a day for 6 months, as supplementary treatment in tuberculosis. A sweet prepared with its baked flowers and sesamum seeds considered useful for cough and cold. Butter like substance that emerges on churning its oil along with that of mustard oil, *Pongamia pinnata* and cow's ghee applied on the chest for 7 days as a remedy for pneumonia and cold.
Active Constituents	:	Polysaccharides PS-AI constituting galactose, arabinose, rhamnose, xylose, glucuronic acid and polysaccharides PS- AII constituting galactose, glucose, arabinose and glucuronic acid isolated from flowers.

References	:	Ambasta (1986), Anonymous (1999), Bhattacharjee (1998), Das (1997), Jain (1968), Jha and Varma (1991), Mishra (2003), Mukherjee and Namhata (1990), Rastogi and Mehrotra (1993), Sharma (2003), Singh and Pandey (1998).

Maerua oblongifolia (Forsk.) A. Rich.

Syn.: *Capparis oblongifolia* Forsk.; *Maerua arenaria* (DC.) Hook.

Family: Capparidaceae

Part/s Used	:	Leaves.
Utilization	:	2 tsp of mixture obtained by pounding its leaves with cumin, onion and cow's milk, given early morning for 5 days, for cough and cold.
References	:	Chopra *et al.* (1956), Upadhyay and Chauhan (2002).

Malva parvifolia Linn.

Family: Malvaceae

H. Names	:	*H.*: Panirak.
Distribution	:	N.W. Himalaya upto 525m, Punjab plains, Mumbai, U.P., Upper Bengal, Deccan, Mysore and Madurai.
Part/s Used	:	Whole Plant and Seeds.
Utilization	:	Seeds used as a demulcent in cough, and whole plant considered useful in treatment of pulmonary infection.
References	:	Ambasta (1986), Das and Agarwal (1991), Kaul (1997), Kirtikar and Basu (1984).

Malva rotundifolia Linn.

Syn.: *Malva vulgaris* Fries.

Family: Malvaceae

Eng. and H. Names	:	*Eng.*: Cheese cake flower, Dwarf Mallow;
		H.: Khubasi, Sonchala.
Distribution	:	Plains of N. India upto 3,048m and Kumaon.
Part/s Used	:	Seeds.
Utilization	:	Seeds used to relieve bronchitis and cough.
References	:	Ambasta (1986), Caius (2003), Das and Agarwal (1991), Kirtikar and Basu (1984), Watt (1972).

Malva sylvestris Linn.

Syn.: *Malva sylvestris* var. *mauritiana* (DC.) Mastero

Family: Malvaceae

Eng. and H. Names	:	*Eng.*: Cheese flower, Common Mallow, Marsh Mallow
		H.: Gulkhari, Kunzi, Vilayati Kangai.
Distribution	:	Temperate W. Himalaya from Kashmir- Kumaon upto 2,400m. Bihar. Parts of Peninsular India.
Part/s Used	:	Whole Plant.
Utilization	:	All parts of plant considered beneficial for sore throat, chronic bronchitis and whooping cough.
Active Constituents	:	These are carotene, calcium, polysaccharides and carbohydrates (plant), anthocyanins (leaves and flowers), and mucin (flowers).
References	:	Ambasta (1986), Thakur *et al.* (1989).

Malva verticillata Linn.

Syn.: *Malva neilgherrensis* Wight; *M. alchemillaceflora* Wall.

Family: Malvaceae

Distribution	:	Temperate Himalaya upto 4,000m.
Part/s Used	:	Roots.
Utilization	:	Used to induce vomiting in whooping cough.
References	:	Ambasta (1986), Kirtikar and Basu (1984), Watt (1972).

Mangifera indica Linn.
[Pl. 3D]

Syn.: *M. domestica* Gartn.

Family: Anacardiaceae

Eng., Sans. and H. Names	:	*Eng.*: Cuckoos Joy, Mango, Spring Tree;
		Sans.: Alipriya, Amra, Manjari, Sripriya;
		H.: Aam, Amb.
Distribution	:	Cultivated throughout tropical India.
Part/s Used	:	Seeds.
Utilization	:	Powdered roasted seed kernels (20-30g) taken for relief in cough and asthma.
Active Constituents	:	Seeds afford stearic and oleic acids (86 per cent), β-sitosterol, (+) catechin, (±) catechin, (-) epicatechin and leucocyanidin.

References	:	Ambasta (1986), Das and Agarwal (1991), Kirtikar and Basu (1984), Kumar (2002), Raju (2000), Rastogi and Mehrotra (1991, 1995), Sharma (2003), Singh and Pandey (1998).

Marrubium vulgare Linn.

Family: Lamiaceae

Eng. and H. Names	:	*Eng.*: Horehound, White Hoarhound;
		H.: Paharigandana.
Distribution	:	Kashmir: 1,700-2,700m.
Part/s Used	:	Bark, Leaves and Stem.
Utilization	:	Infusion of bark recommended for irritation in the bronchia. Powdered leaves combined with honey useful in cough.
Active Constituents	:	These are a volatile oil, gallic acid and a bitter principle marrubin (plant), lactoyl flavonoids -luteolin and apigenin- 7- lactates together with their 2''- O–glucuronides and 2''- O–β–D-glucosides along with apigenin, luteolin, and their 7 glucosides, apigenin -7- (6''- p- coumaroyl) glucoside, vicenin2, vitexin and chrysoeriol (leaves).
References	:	Ambasta (1986), Bhattacharjee (1998), Duke *et al.* (2002), Greenish (1999), Rastogi and Mehrotra (1995), Sharma (2003), Kirtikar and Basu (1984).

Marsilea minuta Linn.

Family: Marsileaceae

Distribution	:	Alwar, Ajmer, Mt. Abu and Sirohi of Rajasthan.
Part/s Used	:	Leaves.
Utilization	:	Decoction of leaves along with ginger used in cough and bronchitis
Active Constituents	:	3- hydroxytriacontan-11-one, protocatechuic, caffeic, p–coumaric, vanillic and ferullic acid isolated from leaves.
References	:	Ambasta (1986), Bhattacharjee (1998), Rastogi and Mehrotra (1993, 1995), Sharma (2003).

Melaleuca leucadendron Linn.

Family: Myrtaceae

Eng. and H. Names	:	*Eng.*: Bottle Brush, Cajuput tree;
		H.: Kayaputi.

Distribution	:	Cultivated in Indian gardens.
Part/s Used	:	Fresh Leaves and Terminal branches.
Utilization	:	Oil extracted from its terminal branches used for curing chronic laryngitis and bronchitis.
Active Constituents	:	Leaf oil affords 1,8-α-cineole (48 per cent), p-cymene (13.2 per cent), α-terpineol (9.8 per cent), limonene (4.8 per cent), α-pinene (3.8), linalool (3.4), β-pinene (2.6 per cent), terpinen–4–ol (1.6 per cent), phellandrene, isopropylanisole, carveol, α- copaene, β-elemene, humulene and ?-cadinene.
References	:	Ambasta (1986), Kirtikar and Basu (1984), Rastogi and Mehrotra (1991, 1995).

Melitotus officinalis Lam.

Family: Fabaceae

Eng. and H. Names	:	*Eng.*: Common melilot, Hart's clover, King's Crown; *H.*: Aspurk.
Distribution	:	Nubra and Ladakh: 3,500-4,350m.
Part/s Used	:	Plant.
Utilization	:	Plant used in preparation of cigarettes for relief from asthma.
Active Constituents	:	Salicyclic, p–hydroxybenzoic, p -hydroxyphenylacetic, vanillic, gentisic, protocatechuic, syringic, p-hydroxyphenyllactic, gallic, caffeic, melilotic, o- and p-coumaric, ferulic and sinapic acid isolated.
References	:	Chopra *et al.* (1980), Kirtikar and Basu (1984), Rastogi and Mehrotra (1995), Viswanathan (1999).

Melissa parviflora Benth.

Family: Lamiaceae

H. Names	:	*H.*: Bililotan.
Distribution	:	Temperate Himalayan region from Uttarakhand to Sikkim and Mishmi and Khasia Hills.
Part/s Used	:	Leaves.
Utilization	:	Leaves useful in treatment of bronchitis and asthma.
References	:	Ambasta (1986), Kirtikar and Basu (1984), Sharma (2003).

Mentha arvensis Linn.
Family: Lamiaceae

Eng. and H. Names	:	*Eng.*: Corn mint, Japanese Mint, Marsh Mint;
		H.: Podina, Pundinah.
Distribution	:	W.Himalaya: 1,700-2,000m.
Part/s Used	:	Leaves.
Utilization	:	Extract of leaves effective in asthma.
References	:	Bhattacharjee (1998), Kirtikar and Basu (1984), Sharma (2003), Sharma and Sood (1997).

Mentha cardiaca Gerard. *ex* Baker
Family: Lamiaceae

Eng. Name	:	*Eng.*: Scotch Mint.
Part/s Used	:	Leaves
Utilization	:	Thin paste of leaves used as a remedy for bronchitis.
Reference	:	Bhattacharjee (1998).

Mesua ferrea Linn.

Syn.: *M. speciosa* Chois.; *M. pedunculata* Weight; *M. coromandelina* Weight; *M. roxburghii* Weight; *M. Salicina, walkeriana* and *pulchella* Planch and Trian; *M. sclerophylla* Thwaites; *M. thwaitesii* Planch and Triana; *M. ferrea* ssp. thwaitesii Vesque; *M. nagana* Gard.

Family: Clusiaceae

Eng., Sans. and H. Names	:	*Eng.*: Ceylon Iron–Wood, Cobras saffron;
		Sans.: Champeya, Hema, Ibhakhya, Nagakeshara;
		H.: Naghas.
Distribution	:	Himalaya, Bengal, Assam, Evergreen rain forest of N. Kanara, Konkan and Forest of W. Ghats.
Part/s Used	:	Flowers, Roots and Bark.
Utilization	:	Infusion of bark, roots and flowers useful in cough and bronchitis.
Active Constituents	:	These are coumarins-mesuferron, mammeisins, mesuanic acid, steroids–β-sitosterol and triterpenoids–guttiferol.
References	:	Ambasta (1986), Anonymous (1999), Bennet (1987), Kirtikar and Basu (1984), Kumar (2002), Sinha and Sinha (2001), Watt (1972).

Michelia champaca Linn.

Syn.: *Michelia rufinervis* DC.; *M. doldsopa* Ham.; *M. aurantiaca* Wall.; *M. rheedii* Weight.

Family: Magnoliaceae

Eng., Sans. and H. Names	:	*Eng.*: Golden champa, Yellow champa;
		Sans.: Anjana, Atigandlika, Bhramaratithi;
		H.: Champ, Champa, Champaca, Champac.
Distribution	:	Wild in the E. Sub–Himalayan tract; Much cultivated in various parts of India.
Part/s Used	:	Flowers.
Utilization	:	Flowers used as an expectorant in cough and bronchitis.
References	:	Jain and Tarafdar (1970), Watt (1972).

Milletia chrysophylla Dunn.

Family: Fabaceae

Part/s Used	:	Bark.
Utilization	:	Decoction of bark given for cough and disease of upper respiratory tract.
Active Constituents	:	Inulin, saponin and oil isolated.
Reference	:	Gill and Nyawuame (1994).

Mimosa lamata Willd.

Family: Fabaceae

Distribution	:	Punjab. C. and S. India.
Part/s Used	:	Leaves.
Utilization	:	Leaf juice given with goat milk to cure bronchitis.
References	:	Rastogi and Mehrotra (1995), Singh and Pandey (1998).

Momordica charantia Linn.

Syn.: *M. humilis* (Cogn.) *C. Jeffrey; M. muricata* DC.; *M. senegalensis* Lamk.; *Cucumis africanus* L.f.

Family: Cucurbitaceae

Eng. and H. Names	:	*Eng.*: Carilla Fruit;
		H.: Karela, Karoli.
Distribution	:	Cultivated throughout India.
Part/s Used	:	Fruits.

Utilization	:	Fruits considered beneficial for asthma and bronchitis.
Active Constituents	:	4 momordicoside G, F_1, F2, momordicosides K and L, charantin, β-sitosterol- glucoside and a stigast-5-25-diene-3β-o-glucoside, mixture of 2 acylglycosylsterols, benzyl alcohol, myrtenol, cis-3-hexenol, trans-2-hexenal, 1-penten-3-ol and cis-2-penten-1-ol identified in fruits.
References	:	Kirtikar and Basu (1984), Rastogi and Mehrotra (1990, 1993, 1995), Singh and Pandey (1980), Watt (1972).

Monarda fistulosa Linn.

Family: Lamiaceae

Eng. Name	:	*Eng.*: Willd Bergamot.
Part/s Used	:	Leaves and Flowers.
Utilization	:	A tea made from leaves and flowers said to cure bronchitis. Also, steam inhalation of vapours resulted by boiling the dried plant considered by Ojibwas for catarrh and bronchial affections.
References	:	Bhattacharjee (1998), Rajan *et al.* (2002).

Monochoria vaginalis Presl. *ex* Kunth.

Syn.: *Pontederia vaginalis* Burm.; *Monochoria vaginalis* var. *plantaginea* (Roxb.) Solms.

Family: Pontederiaceae

H. Name	:	*H.*: Naunka.
Distribution	:	Throughout India.
Part/s Used	:	Leaves. Roots. Bark.
Utilization	:	Juice of roots, leaves and bark taken with sugar for asthma.
References	:	Ambasta (1986), Bennet (1987), Chakraborthy and Hazara (2003), Kirtikar and Basu (1984), Raju (2000).

Morchella esculenta L.

Family: Morchellaceae

Distribution	:	Kashmir, Chamba and Many parts of N. Punjab.
Part/s Used	:	Fruiting body.
Utilization	:	Infusion helps in relieving cough and cold.
References	:	Chopra *et al.* (1956), Singh (1999).

Morinda citrifolia Linn.
Family: Rubiaceae

Eng., Sans. and H. Names	:	*Eng.*: Indian mulberry;
		Sans.: Achchhuka, Ashyuka;
		H.: Al, Ach, Bartundi, Surangi.
Distribution	:	Cultivated largely in India.
Part/s Used	:	Fruits.
Utilization	:	Baked fruits prescribed with ghee to cure asthma.
Active Constituents	:	Main constituents of fruits are asperuloside, glucose, caprylic and caproic acid.
References	:	Ambasta (1986), Das and Agarwal (1991), Kirtikar and Basu (1984), Mishra (2003), Rastogi and Mehrotra (1991).

Moringa oleifera Lam.
Syn.: *M. pterygosperma* Gaertn.
Family: Moringaceae

Eng., Sans. and H. Names	:	*Eng.*: Drum Strike tree;
		Sans.: Akshiba, Bahumula, Chalusha;
		H.: Mungna, Sainjna, Sandna, Segva.
Distribution	:	Cultivated throughout India.
Part/s Used	:	Roots, Flowers, Pods and Stem-bark.
Utilization	:	Pods used as vegetable beneficial in asthma. Decoction of roots, leaves, bark and flowers prescribed in asthma, cough and bronchitis. Crushed stem-bark inhaled for treating cold.
Active Constituents	:	These are benzyl isthiocyanate (roots), alanine, arginine, glycine, serine, threonine, valine, glutamic and asparatic acid (flowers and fruits).
References	:	Bhatt *et al.* (2002), Kirtikar and Basu (1984), Rajwar (1983), Rao and Henry (1996), Rastogi and Mehrotra (1991, 1995).

Mucuna monosperma DC.
Syn.: *Carpopogon monospermum* Roxb.; *Mucuna cristata* Ham.; *M. corymbosa* Grah.; *M. anguina* Wall.; *C. anguineum* Roxb.
Family: Fabaceae

Eng., Sans. and H. Names	:	*Eng.*: Negrobean;

		Sans.: Dadhipushpi, Khatava, Khatavangi;
		H.: Sonogaravi.
Distribution	:	E. Himalaya, Khasia Hills, Assam and Mysore.
Part/s Used	:	Seeds.
Utilization	:	Seeds consumed as vegetable for relief from cough and asthma.
References	:	Ambasta (1986), Das and Agarwal (1991), Karnik *et al.* (1981), Kirtikar and Basu (1984), Raju (2000), Sur (2002), Watt (1972).

Murraya koenigii (L.) Spr.

Syn.: *Bergera koenigii* Linn.

Family: Rutaceae

Eng., Sans. and H. Names	:	*Eng.*: Curry leaf tree;
		Sans.: Kadarya, Kalasaka, Surabhininiba;
		H.: Gandhella, Kathnim, Mitha Neem.
Distribution	:	Throughout India upto 1,500m.
Part/s Used	:	Leaves.
Utilization	:	15ml decoction of leaves in combination with that of *Paederia foetida* (1:1) and pepper given, twice or thrice daily till cure, to treat common cold.
Active Constituents	:	Active principles of leaves are koeinigin, muconicine, mahanimbine, koenimbine koenigicine, cyclomahanimbine, mahanimbidine, mahanine, girinimbine and isomahanimbine.
References	:	Ambasta (1986), Gogoi and Borthakur (2001), Kirtikar and Basu (1984), Rastogi and Mehrotra (1990, 1991, 1993, 1995), Sharma (2003).

Musa sapientum Linn.

Syn.: *Musa paradisiaca* Linn.

Family: Musaceae

Eng., Sans. and H. Names	:	*Eng.*: Adam's fig, Banana, Fig of India;
		Sans.: Ambusara, Balakapriya, Bhamphala, Mocha, Nisara, Rambha, Sukumara;
		H.: Amrit, Kachkula, Kela.
Distribution	:	Cultivated throughout India and tropics.

Part/s Used	:	Flowers and Leaves.
Utilization	:	Both flowers and leaves considered useful for curing asthma and whooping cough.
Active Constituents	:	Principle ingredients of flowers are 9,19-cyclotriterpene ketone, atriterpenoid-24 (R)- 4α,14α,24-trimethyl-5α-cholesta-8,25 (27)-dien-3β-ol, dopamine, dopa, noradrenalin, serotonin and caffeic, cinnamic, p-coumaric, ferullic, gallic and protocatechuic acids, campesterol, β-sitosterol, stigasterol, cyclomusalenol and cyclomusalenone.
References	:	Jain (1984), Kirtikar and Basu (1984), Rastogi and Mehrotra (1993, 1995), Virjee *et al.* (1984).

Mussaenda frondosa Linn.

**Syn.: *M. flavescens* and *M. dorinia* Ham.; *M. formosa* Linn.;
M. villosa Wall.; *M. corymbosa* Roxb; *M. zeylanica* Burm.**

Family: Rubiaceae

H. Name	:	H.: Bedina.
Distribution	:	N. E. Himalaya, Bengal and S. India.
Part/s Used	:	Leaves. Flowers.
Utilization	:	Leaves and flowers considered beneficial in the treatment of cough and asthma.
Active Constituents	:	Ferulic acid, hyperin, quercetin, rutin, sinapic acid and β-sitosterol glucoside isolated from sepals.
References	:	Baruah and Sharma (1987), Mukherjee *et al.* (1986), Rastogi and Mehrotra (1995), Watt (1972).

Mussaenda glabra Linn.

Family: Rubiaceae

Distribution	:	Tropical Himalaya, Assam upto 1,500m.
Part/s Used	:	Roots, Leaves and Flowers.
Utilization	:	Infusion of leaves or chewing it with betel leaf lessens cough. Decoction of roots taken for cough. Flowers considered good for asthma.
References	:	Ambasta (1986), Chopra *et al.* (1980).

Myrica esculenta Buch. Ham. *ex* D. Don

Syn.: *M. forquhariana* Wall.; *M. sapida* Wall.; *M. nagi* Hook. f. in Part, non Thunb.; *M. integrifolia* Roxb.; *M. missianes* Wall; *M. rubra* Sieb. and Zucc; *Nageia japonica* Gaertn; *M. javanica* Blume; *M. longifolia* and *M. lobii* Teysm and Binnend

Family: Myricaceae

Eng. and H. Names	:	*Eng.*: Box myrtle;
		H.: Kaiphal.
Distribution	:	Sub-tropical Himalaya: 1,000- 2,000m.
Part/s Used	:	Bark.
Utilization	:	Decoction of stem-bark useful in chronic cold, cough, flu, asthma, bronchitis and lung affections.
Active Constituents	:	Myricanol glucosides together with myricanol and myricanone, β-sitosterol, taraxerol,myricadiol, triterpene- 28-hydroxy-D-friedoolean-14-en-2-one, taraxerone, epigallocatechin-3-O-gallate, epigallo-catechin (4β→8)-epigallocatechin-3-O-gallate, 3-O-galloylepigallocatechin (4β→8)-epigallocatechin-3-0-gallate, gallic acid isolated from stem-bark.
References	:	Ambasta (1986), Ansari (1991), Kumar (2002), Negi *et al.* (1985), Raju (2000), Rastogi and Mehrotra (1990, 1991, 1993, 1995), Sharma (2003), Watt (1972).

Myrica nagi Thunb.

Family: Myricaceae

H. Name	:	*H.*: Kaiphal.
Distribution	:	Sub-Tropical Himalaya and Khasia Hills.
Part/s Used	:	Bark.
Utilization	:	Decoction of bark effective in cough and asthma.
Active Constituents	:	Chief active principle is myricitrin.
References	:	Chopra *et al.* (1956), Islam (2000).

Myroxylon balsamum (L.) Harms

Syn.: *M. toluiferum* H. B. and K.

Family: Fabaceae

Eng. Name	:	*Eng.*: Tolu Balsam tree.
Part/s Used	:	Volatile oil–Oil of Tolu Balsam.

Utilization	:	Vapours of heated oil inhaled for cure of respiratory aliments, also used as an ingredient of cough mixture.
Active Constituents	:	Oil affords resinous material (75-80 per cent) like cinnamic and benzoic acid.
References	:	Ambasta (1986), Bhattacharjee (1998), Duke *et al.* (2002), Sharma (2003).

Myrtus communis Linn.

Family: Myrtaceae

Eng. and H. Names	:	*Eng.*: Common Myrtle;
		H.: Habulas, Murad, Vilayati mehndi.
Distribution	:	Grown in gardens throughout India.
Part/s Used	:	Leaves and Flowers.
Utilization	:	Leaves and flowers yield essential oil, Oil of Myrtle, applied in affection of respiratory tract.
Active Constituents	:	Active principle of essential oil are lauric, myristic, palmitic, stearic, oleic, linoleic 7 linolenic acid, α-pinene, myrcene, phellandrene, limonene, ?-terpinene, p-cymene, linalool, linalyl acetate, β-caryophylllene, α-terpineol and methyl eugenol.
References	:	Ambasta (1986), Kirtikar and Basu (1984), Rastogi and Mehrotra (1993).

Nardostachys grandiflora DC.

Syn.: *Nardostachys jatamansi* DC.

Family: Valerianaceae

Sans. and H. Names	:	*Sans.*: Jatamansi;
		H.: Jatamansi.
Distribution	:	Alpine zones of C. and E. Himalaya: 3,000-4,500m.
Part/s Used	:	Roots and Leaves.
Utilization	:	Infusion of roots and leaves useful in asthma and bronchitis.
References	:	Aswal (1996), Jain (1968).

Naregamia alata Wight and Arn.

Family: Meliaceae

Eng. and Sans. Names	:	*Eng.*: Goanese ipecacuanha;
		Sans.: Kandalu.

Distribution	:	W. and S. India.
Part/s Used	:	Roots.
Utilization	:	Decoction of roots beneficial for chronic bronchitis.
Reference	:	Ambasta (1986).

Nelumbo nucifera Gaertn.
Syn.: *Nelumbium speciosum* Willd.
Family: Nymphaeaceae

Eng., Sans. and H. Names	:	*Eng.*: Chinese waterlilly, Lotus;
		Sans.: Abja, Ambhoja, Amlana, Jalajanma;
		H.: Ambuj, Kamal, Lalkamal, Padam.
Distribution	:	Occurring throughout warmer parts of the country upto 1,000m.
Part/s Used	:	Stem.
Utilization	:	Extract of stem effective in bronchitis and pneumonia.
Active Constituents	:	Nelumbine.
References	:	Das (1997), Singh and Pandey (1998).

Nepeta cataria Linn.
Family: Lamiaceae

Eng. Name	:	*Eng.*: Catmint.
Distribution	:	Temperate W. Himalaya.
Part/s Used	:	Leaves and Stalks.
Utilization	:	Dried leaves and stalks smoked to relieve respiratory ailments. A tea made from plant good for treating cold.
References	:	Ambasta (1986), Bhattacharjee (1998), Sharma (2003).

Nephrolepis cordifolia (L.) Presl.
Family: Davalliaceae

Part/s Used	:	Roots and Fronds.
Utilization	:	Juice of root (4 tsp) given, thrice a day, for cough and cold. Decoction of fresh fronds also given in cough.
References	:	Ambasta (1986), Manandhar (1996), Rastogi and Mehrotra (1995).

Nigella sativa Linn.

Family: Ranunculaceae

Eng., Sans. and H. Names	:	*Eng.*: Black Cumin, Small fennel;
		Sans.: Sausvi;
		H.: Kalajira, Kalonji.
Distribution	:	An occasional weed of cultivation.
Part/s Used	:	Seeds.
Utilization	:	Essential oil from seeds considered beneficial in cough and bronchial asthma.
References	:	Ambasta (1986).

Ochna jabotapita Linn.

Syn.: *O. squarrosa* Linn.; *O. lucida* Lamk.; *O. refescenns* Thunb.; *O. obtusata* DC.; *O. nitida* Thunb.

Family: Ochnaceae

Eng. and Sans. Names	:	*Eng.*: Golden Champak;
		Sans.: Kanakchampa.
Distribution	:	Assam and W. Peninsula.
Part/s Used	:	Roots.
Utilization	:	Decoction of roots useful for asthma.
References	:	Ambasta (1986), Kirtikar and Basu (1984).

Ochna obtusata DC. var. *pumila* (Buch.- Ham. *ex* DC.) Kanis

Syn.: *O. pumila* Buch.-Ham. *ex* DC.

Family: Ochnaceae

Part/s Used	:	Roots.
Utilization	:	Fresh roots chewed for curing asthma.
References	:	Bennet (1987), Varma (1997).

Ocimum basilicum Linn.

Syn.: *O. pilosum* Willd.; *O. album* Linn.; *O. minimum* Burm.; *O. hispidum* Lamk.; *O. menthaefolium* Benth.; *Plectranthus barrelieri* Spreng.; *O. caryophyllatum* Roxb.

Family: Lamiaceae

Eng., Sans. and H. Names	:	*Eng.*: Basil, Common basil, Sweet Basil;
		Sans.: Munjariki, Surasa, Varvara;
		H.: Babui Tulsi, Bahari, Kalitulsi.

Distribution	:	Cultivated in various parts of India.
Part/s Used	:	Leaves, Twigs and Flowers.
Utilization	:	Considered useful in asthma and cough.
Active Constituents	:	Leaves yield thymol, xanthimicrol, butyl caffeate, p-coumaric acid, aesculetin, eriodictyol, its 7-glucoside and vicenin.
References	:	Ambasta (1986), Jain (1995), Rastogi and Mehrotra (1995), Sharma (2003).

Ocimum gratissimum Linn.
Syn.: *O. citronatum* Ham.; *O. robustum* Heyne.
Family: Lamiaceae

Eng., Sans. and H. Names	:	*Eng.*: Large Basil, Lemon Basil, Shrubby Basil;
		Sans.: Ajaka, Nidralu, Sugandhi, Vriddhutulsi;
		H.: Bantulsi, Malatulsi, Ramtulsi.
Distribution	:	Throughout India and often cultivated in gardens.
Part/s Used	:	Whole Plant.
Utilization	:	Decoction given for cough, cold and bronchitis.
Active Constituents	:	Essential oil from plant contains citral, geraniol, citronellol, geranyl acetate and sesquiterpenes.
Biological Activity	:	Aqueous extract exhibits a complete inhibition of the growth of all the 3 tested *Mycobacterium tuberculosis* strains. Antiasthmatic activity found positive in Guinea pig.
References	:	Ambasta (1986), Gill *et al.* (1993), Jain *et al.* (1995), Kirtikar and Basu (1984), Thakur *et al.* (1989), Watt (1972).

Ocimum sanctum Linn.
[Pl. 4A]
Syn.: *O. monachorum* Linn.; *O. tenuiflorum* Linn.; *O. inodorum* Burm.
Family: Lamiaceae

Eng., Sans. and H. Names	:	*Eng.*: Holy Basil, Sacred Basil;
		Sans.: Ajaka, Manjari, Shyama, Subhaga;
		H.: Baranda, Kalalulsi, Tulsi, Varanda.
Distribution	:	Ascending upto 6,000m in Himalaya.
Part/s Used	:	Leaves.

Utilization	:	Fresh leaf juice given in bronchitis and cough. 2 pills each of 5g prepared by pounding its fresh leaves with the rhizome of *Zingiber officinale* (1:1) given, twice daily for 5 days, to cure cough, cold and flu.
Active Constituents	:	Leaves contain 0.075 per cent oil, β-carotene, and ursolic acid.
References	:	Ambasta (1986), Bhattacharjee (1998), Bhatt *et al.* (2001), Caius (2003), Gogoi and Borthakur (2001), Jain (1968), Kirtikar and Basu (1984), Mandal and Basu (1996), Panthi and Chaudhary (2003), Rastogi and Mehrotra (1995), Sharma (2003), Sharma and Sood (1997), Shome *et al.* (1996), Watt (1972).

Oldenlandia corymbosa Linn.

Syn.: *O. ramosa* Roxb.; *Hedyotis corymbosa* (L). Lam.

Family: Rubiaceae

Sans. and H. Names	:	*Sans.*: Parpata;
		H.: Dhaman Papar.
Distribution	:	Throughout India.
Part/s Used	:	Plant.
Utilization	:	Extract of plant employed for curing bronchitis and pneumonia.
Active Constituents	:	10-acyliridoid glucosides, asperulosides, deacetyl asperuloside, asperulosidic acid, deacetyl asperulosidic and scandoside methyl ester isolated from the plant.
References	:	Kirtikar and Basu (1984), Rastogi and Mehrotra (1990, 1993, 1995).

Oldenlandia umbellata (Linn.) Lam.

Family: Rubiaceae

Eng. and H. Names	:	*Eng.*: Chaya root, Indian Madder;
		H.: Chirval.
Distribution	:	Orissa, Bengal and Deccan.
Part/s Used	:	Leaves and Roots.
Utilization	:	Leaves and root of this plant considered to possess strong expectorant property and prescribed in cases of bronchitis and asthma.
Active Constituent	:	Auricularine isolated.

References	:	Chopra *et al.* (1956), Kirtikar and Basu (1984), Maya *et al.* (2003).

Oligochaeta ramosa (Roxb.) Wagenitz.

Syn.: *Volutarella ramosa* (Roxb.) Santapu; *V. divaricata* Benth. and Hook. f.; *Tricholepis procumbens* Wight.

Family: Asteraceae

H. Name	:	*H.*: Badavard.
Part/s Used	:	Stem and Leaves.
Utilization	:	2g paste of stem and leaves consumed, twice a day for 4 days, for cough and cold.
Active Constituents	:	Jaccosidin and cyanopicrin isolated from aerial parts.
References	:	Ambasta (1986), Rastogi and Mehrotra (1993), Vishwanathan *et al.* (2001).

Ophioglossum vulgatum Linn.

Family: Ophioglossaceae

Eng. Names	:	*Eng.*: Adder's Tongue, Common Tongue fern.
Distribution	:	Sikkim 1,350m.
Part/s Used	:	Rhizome.
Utilization	:	Decoction of rhizome beneficial for bronchial pulmonary diseases.
Active Constituents	:	Flavonol triglycoside-3o-methylquercetin-7-O-diglucoside-4'-O-glucoside isolated.
References	:	Nwosu (2001), Rastogi and Mehrotra (1990).

Opuntia tuna Mill.

Family: Cactaceae

Part/s Used	:	Flat branches. Fruits.
Utilization	:	Flat branches and fruits used to cure asthma.
Reference	:	Bhattacharjee (1998).

Opuntia dillenii Haw.

Syn.: *Cactus indicus* Roxb.; *Cactus dillenii* Ker-Gawler; *Opuntia stricta* (Haw) Haw. var *dillenii* Benson.

Family: Cactaceae

Eng., Sans. and H. Names	:	*Eng.*: Prickly Pear, Slipper Thorn;
		Sans.: Guda, Nagaphana, Netrari;
		H.: Nagphana, Nagaphani.

Distribution	:	Introduced in India.
Part/s Used	:	Fruits.
Utilization	:	Baked fruits given for whooping cough in Deccan. Also, syrup of fruits given to control spasmodic cough.
Active Constituents	:	Glycoside of isorhamnetin, quercetin, isoquercetin and narcissin isolated from flower and fruits.
References	:	Chopra *et al.* (1980), Kirtikar and Basu (1984), Rastogi and Mehrotra (1990), Watt (1972).

Origanum vulgare Linn.

Syn.: *O. normale* Don; *O. laxiflora* Royle.

Family: Lamiaceae

Eng. and H. Names	:	*Eng.*: Common or Wild Marjoram, Eng Marjoram;
		H.: Mirzanjosh, Sathra.
Distribution	:	Temperate Himalaya from Kashmir to Sikkim: 2,350–4,000m.
Part/s Used	:	Whole Plant.
Utilization	:	Plant infusion useful in asthma.
Active Constituents	:	These are proteins, fats, crude fiber, carbohydrates, minerals and vit. A, B and C (leaves), an antioxidative glucoside and phenolic acid–2–caffeoyloxy–3–[2–(4 -hydroxybenzyl)–4, 5 -dihydroxyl] phenylpropionic acid (I) (plant).
References	:	Ambasta (1986), Bhattacharjee (1998), Kirtikar and Basu (1984), Rastogi and Mehrotra (1995), Watt (1972).

Osbeckia zeylanica L.f.

Family: Melastomataceae

Part/s Used	:	Whole Plant.
Utilization	:	Half a tsp of plant paste consumed with milk, once a day for 3 days, for cough and cold.
Reference	:	Vishwanathan *et al.* (2001).

Oxalis corniculata Linn.

Syn.: *O. repens* Thumb.; *O. pusilla* Salisb.

Family: Oxalidaceae

Eng., Sans. and H. Names	:	*Eng.*: Indian Sorrel, Yellow Oxalis;
		Sans.: Amlika, Amlotaja, Chukrita;
		H.: Amrul, Anbotic, Chalmori.

Distribution	:	Throughout the warmer regions of India.
Part/s Used	:	Leaves.
Utilization	:	Leaf juice taken along with honey as a remedy for cough.
References	:	Kirtikar and Basu (1984), Kumar (2002), Watt (1972).

Paederia foetida Linn.
Syn.: *P.ovata* Miq, *P sessiliflora* DC.
Family: Rubiaceae

Sans. and H. Names	:	*Sans.*: Balya, Gandhali, Prabal, Sarani;
		H.: Bakuchi, Gandha Prasarini, Gundali, Krip.
Distribution	:	Most parts of the country and Himalayas upto1,700m.
Part/s Used	:	Leaves.
Utilization	:	15ml decoction of leaves of *P. foetida* and *Murraya koenigii* in equal quantity with pepper, taken twice or thrice daily.
Active Constituents	:	Hentriacontane, hentriacontanol, methyl mercapten, ceryl alcohol, palmitic acid, sitosterol, stigasterol, campesterol, ursolic acid and iridoid glycosides-asperuloside, paederoside and scandoside isolated from leaves.
References	:	Anonymous (1999), Bhattacharjee (1998), Kirtikar and Basu (1984), Rastogi and Mehrotra (1991, 1995), Watt (1972).

Pandanus fascicularis Lamk.
Syn.: *Pandanus tectorius* Soland *ex* Parkinson; *P. odoratissimus* Linn.f.
Family: Pandanaceae

Eng., Sans. and H. Names	:	*Eng.*: Screw Pine, Umbrella Tree;
		Sans.: Dalapushpa, Indukalika, Jambuka, Malina;
		H.: Gagandhul, Keora, Ketgi, Keura.
Distribution	:	Sea coast of the Indian Peninsula on both sides of Andaman.
Part/s Used	:	Leaves.
Utilization	:	Basal portion of leaf mixed with salt is said to be useful in treating cough.
Active Constituents	:	β-sitosterol, β-sitostenone, stigmasterol and stigast-4-en-3, 6-dione isolated.

References	:	Girach *et al.* (1997), Kirtikar and Basu (1984), Rastogi and Mehrotra (1995).

Paris polyphylla Smith
Family: Liliaceae

Distribution	:	Temperate Himalaya: 1,500-3,000m.
Part/s Used	:	Rhizome.
Utilization	:	Beneficial for asthma.
Active Constituents	:	Polyphyllins A-H isolated.
References	:	Rana *et al.* (1996), Rastogi and Mehrotra (1990, 1995).

Passiflora foetida Linn.
Family: Passifloraceae

Eng. and H. Names	:	*Eng.*: Stinking Passion Flower; *H.*: Mukhopeera.
Distribution	:	Widely distributed in India.
Part/s Used	:	Fruits.
Utilization	:	Decoction of fruits used in asthma.
Active Constituents	:	Harman.
References	:	Ambasta (1986), Das and Agarwal (1991).

Pedalium murex Linn.
Family: Pedaliaceae

Sans. and H. Names	:	*Sans.*: Gokshura, Tittagokshura; *H.*: Baragokhru, Faridbute.
Distribution	:	Gujarat, Konkan and Deccan Peninsula.
Part/s Used	:	Plant.
Utilization	:	Extract of plant effective in bronchitis and pneumonia.
References	:	Das (1997), Kirtikar and Basu (1984).

Peganium harmala Linn.
Family: Rutaceae

Eng. and H. Names	:	*Eng.*: Foreign Heena, Harmal, Wild Rue; *H.*: Harmal, Kaladana, Lahouri-Harma.
Distribution	:	Widely distributed in N.W India.
Part/s Used	:	Seeds.

Utilization	:	Dried seeds considered effective in chronic bronchitis.
Active Constituents	:	Principle alkaloids of seeds are harmine, harmaline, harmalol and peganine.
References	:	Ambasta (1986), Arshad *et al.* (1997), Bhattacharjee (2001), Kirtikar and Basu (1984), Rastogi and Mehrotra (1990), Sharma (2003).

Pellalaeo calomelanos Link.

Syn.: *Pitygrogramma calomelanos* (L.) Link.; *Pteris calomelanos* (L.) Bedd.

Family: Polypodiaceae

Distribution	:	Dehradun and Kumaon ascending upto 1,700m.
Part/s Used	:	Fronds.
Utilization	:	Fronds smoked for relief from asthma and cold.
References	:	Ambasta (1986), Chopra *et al.* (1956), Rastogi and Mehrotra (1995).

Pergularia daemia (Forsk.) Chiov.

Syn.: *P. extennsa* N.E.Br.; *Daemia extensa* R.Br.; *Asclepias daemia* Forsk.

Family: Asclepiadaceae

Sans. and H. Names	:	*Sans.*: Phala kantaka, Vishanika, Yugaphala;
		H.: Jutuk, Sagovani, Utran, Utranajutuka.
Distribution	:	All parts of India upto 1,500m.
Part/s Used	:	Leaves.
Utilization	:	Leaves used as an emetic in bronchial congestion.
Active Constituents	:	Triterpene-3 β-hydroxyfriedelan-7-one; lupeol, lupeol acetate, oleanolic acid, putranjivadione and β-sitosterol isolated from leaves.
References	:	Ambasta (1986), Bhatt *et al.* (1999), Bhattacharjee (1998), Kirtikar and Basu (1984), Kshirsagar *et al.* (2003), Rastogi and Mehrotra (1993), Sharma (2003), Singh and Pandey (1986).

Pericampylus glaucus (Lam.) Merrill

Syn.: *P. incanus* Miers.; *Menispermum glaucum* Lamk.

Family: Menispermaceae

H. Name	:	*H.*: Barakkanta.
Distribution	:	Assam, Khasia and Sikkim.
Part/s Used	:	Leaves.

Utilization	:	Infusion of leaves given for asthma.
References	:	Ambasta (1986), Chopra *et al.* (1956).

Phaseolus mungo Linn.
Syn.: *Vigna mungo* (L.) Hepper
Family: Fabaceae

Eng., Sans. and H. Names	:	*Eng.*: Green gram, Mung;
		Sans.: Buktiprada, Suphala,Varnarha;
		H.: Mung, Munj.
Distribution	:	Vastly cultivated in India.
Part/s Used	:	Seeds.
Utilization	:	Recommended in bronchitis.
References	:	Kirtikar and Basu (1984), Rastogi and Mehrotra (1991, 1995).

Phlogacanthus thyrsiflorus Wall.
Family: Acanthaceae

Distribution	:	Upper Gangetic plains, Sub-tropical Himalayas upto 1,350m. Chota Nagpur.
Part/s Used	:	Leaves.
Utilization	:	Decoction of leaves taken with honey for cough.
References	:	Chopra *et al.* (1956), Singh *et al.* (1997).

Phoenix dactylifera Linn.
Family: Arecaceae

Eng., Sans. and H. Names	:	*Eng.*: Arabian date, Cultivated date palm, Date palm;
		Sans.: Dipya, Mudarika, Phalapushpa, Rajajambu;
		H.: Khaji, Khajur.
Distribution	:	Cultivated and self grown in dry districts.
Part/s Used	:	Fruits.
Utilization	:	Dates boiled in milk for 5min prescribed for asthma, cough, cold, bronchitis and tuberculosis (4-5 dates, once daily).
Active Constituents	:	Fruits contain vit A, B and D.
References	:	Ambasta (1986), Caius (2003), Chopra *et al.* (1956), Das and Agarwal (1991), Kirtikar and Basu (1984), Sen and Batra (1997).

Phoenix pusilla Gaertn.
Syn.: *P. farinifera* Roxb.
Family: Arecaceae

Distribution	:	Coromandel Coast.
Part/s Used	:	Fruits.
Utilization	:	Fruits beneficial for cough and asthma.
References	:	Caius (2003), Kirtikar and Basu (1984).

Phyllanthus amarus Schum. and Thonn.
Syn.: *P. niruri* auct. non L.; *P. nanus* Hook.
Family: Euphorbiaceae

H. Name	:	*H*.: Bhonyabali, Jangali amla, Jaramala.
Distribution	:	A tropical weed.
Part/s Used	:	Fruits.
Utilization	:	10 fruits consumed in the morning for a month to cure asthma.
Active Constituents	:	4-methoxysecurinine and 4-methoxynorsecurinine reported from aerial parts of plant.
References	:	Thakur *et al.* (1989), Upadhyay and Chauhan (2002).

Phyllanthus urinaria Linn.
Syn.: *P. leprocarpus* Wight; *P. alatus* Blume; *P. echinatus* Herb.
Family: Euphorbiaceae

Sans. and H. Names	:	*Sans*.:Adhyanda, Aphala, Chorata, Jharika, Nilolika, Tamalika;
		H.: Hazarmani, Lalbhuinanvalah.
Distribution	:	Throughout India from Punjab–Assam.
Part/s Used	:	Leaves and Fruits.
Utilization	:	Extract of leaves given, twice a day, for cough. Fruits also useful in asthma and bronchitis.
References	:	Caius (2003), Jha and Varma (1996), Kirtikar and Basu (1984), Watt (1987).

Picrorrhiza kurrooa Royle
Syn.: *P. scrophulariiflora* Pennell.
Family: Scrophulariaceae

Eng., Sans. and H. Names	:	*Eng*.: Hellebore;

		Sans.: Katuka;
		H.: Kuti.
Distribution	:	Alpine Himalayas from Kashmir to Sikkim: 3,000-5,000m.
Part/s Used	:	Roots.
Utilization	:	1 tsp root powder given, 3 times a day, with honey for asthma.
Active Constituents	:	Picrorhizin; steroids, beta sitosterol and kutkisterol, vanillic and cathartic acids isolated.
References	:	Arya and Prakash (1999), Jain (1968), Sharma and Sood (1997), Sharma (2003), Singh (1996), Smith (2000).

Piliostiga reticulatum (DC.) Hochst.
Family: Fabaceae

Part/s Used	:	Roots. Stem.
Utilization	:	Decoction of root and stem given for cough.
Reference	:	Gill and Nyawuame (1994).

Pimpinella anisum Linn.
Syn.: *Anisum vulgare* Gaertn.
Family: Apiaceae

Eng., Sans. and H. Names	:	*Eng.*: Aniseed, Spanish Aniseed Oil;
		Sans.: Shetapushapa;
		H.: Badian, Saunf.
Distribution	:	Commercially cultivated as a spice crop in various parts of India.
Part/s Used	:	Fruits.
Utilization	:	Powdered dried fruits used for relief from asthma. Oil from fruits antispasmodic and expectorant in bronchial catarrh.
Active Constituents	:	Volatile oil affords luteolin (0.125 per cent), luteolin-7-O-glucosides (0.128 per cent) and luteolin-7-O-xyloside (0.158 per cent).
References	:	Ambasta (1986), Duke *et al.* (2002), Harborne and Baxter (2001), Rastogi and Mehrotra (1984).

Pinus palustris Mill.

Family: Pinaceae

Eng. Names	:	*Eng.*: Long leaf pine, Pitchpine.
Part/s Used	:	Gum and Pinetar.
Utilization	:	Recommended for bronchitis and tuberculosis.
Active Constituents	:	Natural oleoresins obtained from plant contains 66 per cent resin acid, 25 per cent turpentine, 7 per cent non volatiles and 2 per cent water.
Reference	:	Bhattacharjee (1998).

Pinus roxburghii Sargent

Syn.: *P. longifolia* Roxb.

Family: Pinaceae

Eng., Sans. and H. Names	:	*Eng.*: Chirpine, Himalayan long leaved pine;
		Sans.: Sarala;
		H.: Chil, Chir.
Distribution	:	Himalayan region upto 2,100m.
Part/s Used	:	Turpentine Oil.
Utilization	:	Inhaling the vapours of turpentine considered useful in chronic bronchitis.
Active Constituents	:	Oleo-resin yields turpentine oil and resin. β -pinene and car-3-ene isolated from turpentine oil.
References	:	Ambasta (1986), Bhattacharjae (1998), Jain (1968), Rastogi and Mehrotra (1984), Sharma (2003).

Pinus sylvestris Linn.

Family: Pinaceae

Eng. Name	:	*Eng.*: Pine.
Distribution	:	Hilly parts of India.
Part/s Used	:	Buds.
Utilization	:	Medicines prepared from buds for chronic catarrh in bronchia.
Active Constituents	:	Young branches and needle yield 0.25-0.35 per cent volatile oil with balsamic odour; pinene, limonene, careen, camphene and sesquiterpene alcohols.
References	:	Bhattacharjee (1998), Sharma (2003).

Piper betle Linn.

Syn.: *Chavica betle* Miq.; *C. chuvya* Miq.; *C. siriboa* Miq.; *P. siriboa* Linn.; *P. peepuloides* Wall; *P. chavya* Ham.

Family: Piperaceae

Eng., Sans. and H. Names	:	*Eng.*: Betel leaf, Betel pepper;
		Sans.: Bhujangalata, Divabhishta, Kalaskanda;
		H.: Pan, Tambuli.
Distribution	:	Cultivated in India in hotter and damper parts.
Part/s Used	:	Petiole.
Utilization	:	A petiole grounded with a pinch of turmeric powder given, once a day, for whooping cough.
Active Constituents	:	Drug yields 0.2-1 per cent of volatile oil, principal constituent of which is betel-phenol. chavieol and cadinene also present in oil.
References	:	Chopra *et al.* (1956), Greenish (1999), Kirtikar and Basu (1984), Sen and Batra (1997), Shome *et al.* (1996), Watt (1972).

Piper cubeba Linn.

Syn.: *Cubeba officinalis* Miq.

Family: Piperaceae

Eng. and Sans. Names	:	*Eng.*: Tailed Pepper;
		Sans.: Kabab Chini.
Distribution	:	Distributed in E. India.
Part/s Used	:	Seeds.
Utilization	:	Dried seeds of unripe fruits useful in chronic bronchitis.
Active Constituents	:	10-18 per cent volatile oil affords cubebin, cubebic acid, 1, 4 cineol, terpene and sesquiterpene.
Reference	:	Bhattacharjee (1998).

Piper longum Linn.

Syn.: *Chavica roxburghii* Miq.; *C. sarmentosa* Miq.; *P. sarmentosum* Wall.; *P. latifolium* Hunter

Family: Piperaceae

Eng., Sans. and H. Names	:	*Eng.*: Long pepper;
		Sans.: Chanchala, Chapala, Goramika, Parkari;
		H.: Gazpipal, Pipal, Piplannrol, Pipulmul.

Distribution	:	C. Himalaya, Khasia and Mikir Hills, Western Ghats and Lower hills of W. Bengal.
Part/s Used	:	Fruits, Seeds and Roots.
Utilization	:	For cough, asthma and other respiratory diseases, powdered fruiting spikes along with leaf juice of *Justicia adhatoda* and honey are taken. 2 pills (3g) each prescribed, twice daily, of mixture of its dried fruits and those of *P. nigrum and Zanthoxylum rhetsa* (1:1:1) in asthma. Extract of rooots administered orally for asthma. Extract of seeds grounded with roots of *Hemidesmus indicus* and *sida spinosa* considered efficacious in asthmatic complaints. Dried unripe fruits (decoction) also useful in cough and chronic bronchitis.
Active Constituents	:	Main constituents are piperine and piplartine, dihydrostigasterol, N-hexadedecane, essential oil (fruits), piperlongumine, piperlonguminine, piperine and sesamum methyl 3, 4, 5-trimethoxycinnamate (roots).
References	:	Balasubramanian and Prasad (1996), Bhattacharjee (1998), Das and Agrawal (1991), Gogoi and Borthakur (2001), Jain (1968), Kirtikar and Basu (1984), Rao and Henry (1996), Rastogi and Mehrotra (1990, 1991), Sharma (2003), Thakur *et al.* (1989), Watt (1972).

Piper nigrum Linn.
[Pl. 4B]

**Syn.: *Piper trioicum* Roxb.; *P. nigrum* var. *trioicum* DC.;
P. malabarense DC.; *P. baccatum* DC.;
Muldera multinervis Miq.; *M. wighiana* Miq.**

Family: Piperaceae

Eng., Sans. and H. Names	:	*Eng.*: Black pepper, Common pepper;
		Sans.: Kolaka, Krishna, Mrishta, Ruksha, Shudha;
		H.: Golmirch, Kalimirch, Safed mirch.
Distribution	:	Hot and moist parts of S. India.
Part/s Used	:	Dried fruits, Seeds and Leaves.
Utilization	:	Decoction of fruits alone or mixed with the bark of *Myristica fragrans* (Jaiphal) taken orally as a remedy for cough and cold. Leaf juice considered good for cough and cold by Konda Doras and Nuka Doras. Paste of dry fruits of *Piper longum, P. nigrum* and *Zanthoxylum rhetsa* in equal quantity with a little water made into pills of

about 3g each and 2 pills (twice daily) prescribed till cure.

Active Constituents	:	Pipercide, n-trans-feruloyltyramine, guineensine, n-isobutyl–2E, 4E, 8Z–Eicosatrienamide, n–isobutyl–2E, 4E–octadecadienamide, pellitorine, phenolic amides–n–trans–feruloyl–piperidine, feruperine and dihydroferuperine isolated from fruits.
References	:	Gogoi and Borthakur (2001), Henry and Rao (1996), Kirtikar and Basu (1984), Rastogi and Mehrotra (1991, 1993), Sen and Batra (1997), Sharma (2003), Singh and Pandey (1998), Watt (1972).

Pistacia integerrima Stewart *ex* Brandis

Syn.: *Rhus integerrima* Wall;
P. chinensis Bunge ssp. *integerrima* (Stewart) Rech.f.

Family: Anacardiaceae

Sans. and H. Names	:	*Sans.*: Chakra, Chakrangi;
		H.: Kakra.
Distribution	:	Hot slopes of W.Himalaya: 700-2,700m.
Part/s Used	:	Leaf Galls.
Utilization	:	Gall on leaves considered useful in cough, asthma and chronic bronchitis.
Active Constituents	:	Gall contains essential oil, 3 tetracyclic triterpenoids - Pistacigerrimones A, B, C, D, E and F.
References	:	Ambasta (1986), Raju (2000), Rastogi and Mehrotra (1995), Watt (1972).

Pistacia khinjuk Stocks

Syn.: *P. integerrima* auct. non Stewart; *Rhus kakrasingee* Royle

Family: Anacardiaceae

Sans. and H. Names	:	*Sans.*: Chandraspada, Karkati, Karkatsmgi;
		H.: Kakra.
Distribution	:	Outer ranges of N.W. Himalaya.
Part/s Used	:	Insect galls on the leaves, petiole and branches. Fruits.
Utilization	:	Powdered burnt fruits and galls mixed with honey good for asthma, phthisis, cough and other conditions of respiratory tract.
Reference	:	Singh (1999).

Pittosporum floribundum Wight and Arn.

Syn.: *P. nepaulense* Rehder and F. H. Wils; *Celastrus verticillata* Roxb.; *Senacia nepaulensis* DC.

Family: Pittosporaceae

Distribution	:	Sub-tropical Himalaya from Sikkim to Garhwal ascending to 1,700m.
Part/s Used	:	Bark.
Utilization	:	A pinch of powdered dried bark given for chronic bronchitis.
Active Constituents	:	Bitter glucoside, essential oil.
References	:	Ambasta (1986), Caius (2003), Chopra *et al.* (1956), Kirtikar and Basu (1984), Watt (1972).

Plantago lanceolata Linn.

Syn.: *P. attenuata* Wall.

Family: Plantaginaceae

Eng. and H. Names	:	*Eng.*: Fighting cocks, Mardheads, Ripple grass;
		H.: Baltanga.
Distribution	:	W. Himalaya from Kashmir to Shimla.
Part/s Used	:	Leaves and Roots.
Utilization	:	Recommended for cough, asthma and pulmonary diseases.
Active Constituents	:	p-hydroxybenzoic, protocatechuic, gentisic, vanillic, syringic, p-coumaric, caffeic, ferullic and p-hydroxyphenylacetic acid, agnuside and aucubin found in leaves.
References	:	Bhattacharjee (1998), Kirtikar and Basu (1984), Rastogi and Mehrotra (1990, 1991, 1995), Sharma (2003), Watt (1972).

Plantago ovata Forsk.

Syn.: *P. decumbens* Forsk.; *P. ispaghul* Roxb.; *P. lanata* Wall.

Family: Plantaginaceae

Eng., Sans. and H. Names	:	*Eng.*: Spogel seeds;
		Sans.: Ishadgola, Snigdhabija, Snigdhajiraka;
		H.: Asli ispafgul, Isabghul, Issufgul.
Distribution	:	Punjab plains.

Part/s Used	:	Seeds.
Utilization	:	Decoction of seeds prescribed in cough and cold.
Active Constituents	:	Seeds contain a semi drying fatty oil, small amounts of aucubin and tannin, an active principle exhibiting acetylcholine like action, no. of amino acids-DL-valine, DL-alanine, glycine, DL-norleucine, L-lysine, glutamic acid, L-cystine, tyrosine, L- (-)–asparagines and polysaccharides.
References	:	Das and Agarwal (1991), Kitikar and Basu (1984), Sinha and Sinha (2001), Thakur *et al.* (1989), Watt (1972).

Plectranthus mollis (Ait.) Spreng.

Syn.: *Ocimum molle* Aiton.; *P. incanus* Link.

Family: Lamiaceae

Distribution	:	Himalayas, upto 2,000m. Khasia Hills. Hills of C. India and W. Ghats.
Part/s Used	:	Seeds.
Utilization	:	Powdered fried seeds given, 4 times a day for 3 days, to cure cough and cold.
Active Constituent	:	Sitosterol isolated from the plant.
References	:	Bennet (1987), Rastogi and Mehrotra (1990, 1991), Varma (1997).

Podophyllum hexandrum Royle

Syn.: *P. emodi* Wall. *ex* Hook. f.; *P. hexandrum* var. *jaeschkei* (Chatt. *et* Muker) Browicz.; *P. emodi* var. *hexandrum* (Royle) Chatt. *et* Muker.

Family: Podophyllaceae

Eng. and H. Names	:	*Eng.*: Indian podophyllum;
		H.: Bakrachimaka, Bhavanbakra, Papra, Papri.
Distribution	:	Inner range of Himalaya from Kashmir to Sikkim: 3,000-4,200m.
Part/s Used	:	Fruits.
Utilization	:	2-3g of pulverized fruits (thrice a day for 10-15 days), prescribed for cough and tuberculosis.
Active Constituents	:	Podophyllotoxin, picropodophyllin, quercetin, podophyllotoxin-β-D-glucoside.
References	:	Sharma and Sood (1997), Sood *et al.* (2001).

Pogostemon heyneanthus Benth.

Syn.: *P. patchouli* Hook. f., non Pelletier; *Origanum indicum* Roth.

Family: Lamiaceae

H. Name	:	*H.*: Pacholi, Peholi.
Distribution	:	Wild and cultivated in W. India from Bombay S.wards.
Part/s Used	:	Leaves.
Utilization	:	Decoction taken for relief in cough and asthma.
References	:	Ambasta (1986), Watt (1972).

Polyalthia rufescens Hook. f. and Thoms.

Family: Annonaceae

Part/s Used	:	Leaves.
Utilization	:	Half a spoon of leaf paste consumed with water (twice a day for 2 days) for cough.
Reference	:	Vishwanathan *et al.* (2001).

Polygala chinensis Linn.

Syn.: *P. telephioides* Willd; *P. brachystachya* DC.non Blume; *P. chinensis* auctt. non L.; *P. arvensis* Willd; *P. prostrata* Willd; *P. rothiana* W. & A.; *P. tranquebarica* Mart.; *P. glaucoides* Wight; *P. glandiflora* Hb. Wight.

Family: Polygalaceae

Eng. and H. Names	:	*Eng.*: Common Indian Milk wort;
		H.: Meradu.
Distribution	:	Throughout India upto 1,700m.
Part/s Used	:	Leaves and Roots.
Utilization	:	Infusion of leaves efficacious in asthma, chronic bronchitis and catarrhal affections. 10g root powder given, twice a day for a month, to cure asthma. Also, paste of its roots with 2 ½ seeds of *P. nigrum* given with cow's hot milk for cough and asthma (twice a day for 7 days).
Active Constituents	:	Arctigenin–4'- glucoside, kaemferol–3 -rhamnoside, myricetin–3-rhamnoside, arctigenin-4'- gentiobiocide and quercetin–3–rutinoside isolated from roots.
References	:	Aditya and Ghosh (1988), Ambasta (1986), Kirtikar and Basu (1984), Rastogi and Mehrotra (1991), Shukla and Verma (1996), Watt (1972).

Polygala glomerata Lour.
Family: Polygalaceae

Distribution	:	Sikkim, Assam and Khasia Hills.
Part/s Used	:	Shoots.
Utilization	:	Infusion effective for asthma and chronic bronchitis.
References	:	Ambasta (1986), Kirtikar and Basu (1984).

Polygala senega Linn.
Family: Polygalaceae

Eng. Name	:	*Eng.*: Rattle snake root.
Part/s Used	:	Roots.
Utilization	:	Employed in pulmonary infection and asthma.
Active Constituents	:	Roots contain triterpenoid, saponins, polygalitol and sengin.
References	:	Ambasta (1986), Bhattacharjee (1998), Greenish (1999).

Polygala sibirica Linn.
Syn.: *P. heyneana* Wight and Arn.
Family: Polygalaceae

Eng. Names	:	*Eng.*: Common milkwort, Japanese senega.
Distribution	:	Temperate and Sub-tropical Himalaya: 350-2,000m.
Part/s Used	:	Roots.
Utilization	:	Decoction of roots given as expectorant in cough, cold and bronchitis.
References	:	Ambasta (1986), Caius (2003), Kirtikar and Basu (1984).

Polygonum amplexicaule Don
Family: Polygonaceae

Distribution	:	Himalayas from Kashmir to Sikkim: 1,800-4,800m.
Part/s Used	:	Roots.
Utilization	:	Deccoction prepared from 1-2g of fresh roots used for cough and cold.
References	:	Arya and Prakash (1999), Rastogi and Mehrotra (1995).

Polygonum plebium R.Br.
Family: Polygonaceae

H. Name	:	H.: Macheti.
Distribution	:	Tropical India and Himalayas upto 2,350m.
Part/s Used	:	Whole Plant.
Utilization	:	Powdered dried plant efficacious in pneumonia.
Active Constituents	:	Oleanolic, betulinic acid, epi-friedelinol, β-sitosterol, quercetin-3-arabinoside and quercetin-3-rutinoside isolated from flowers.
References	:	Ambasta (1986), Chopra *et al.* (1956), Rastogi and Mehrotra (1991).

Pongamia pinnata (Linn.) Merr.
Family: Fabaceae

Eng., Sans. and H. Names	:	*Eng.*: Indian Beach, Karanj, Pongam oil tree;
		Sans.: Karanj, Karanja;
		H.: Karanga.
Distribution	:	All over India near river banks and sea coast.
Part/s Used	:	Seeds.
Utilization	:	Powdered seeds beneficial in bronchitis and whooping cough.
Active Constituents	:	Seeds contain karanjin, pongamol and glabrin.
References	:	Ambasta (1986), Das and Agarwal (1991), Pullaiah *et al.* (1996).

Populus balsamifera Linn.
Syn.: *P. suaveolens* Loud; *P. laurifolia* Ledeb.
Family: Salicaceae

Distribution	:	Dry regions of India.
Part/s Used	:	Bark.
Utilization	:	Tincture prepared from the tree bark useful for treating pulmonary ailments.
References	:	Bhattacharjee (1998), Sharma (2003), Watt (1972).

Portulaca oleracea Linn.

Syn.: *P. laevis* Ham.; *P. suffruticosa* Thw.

Family: Portulacaceae

Eng., Sans. and H. Names	:	*Eng.*: Common Purslane, Garden Purslane, Pot Purslane;
		Sans.: Brihalloni, Lonamla, Lonika;
		H.: Khulfa, Khursa, Lonia, Lunuk.
Distribution	:	Throughout India, upto 1,700m in Himalaya.
Part/s Used	:	Leaves and Flowers.
Utilization	:	Paste of leaves and flowers used for pneumonia, asthma and bronchitis.
Active Constituents	:	Carbohydrates, lipids, glycosides, alkaloids, sterols, triterpenes and flavanoids; oleracins I and II, acylated betacyanins and mucilage (0.42 per cent) isolated from leaves.
References	:	Ambasta (1986), Gill *et al.* (1997), Kirtikar and Basu (1984), Rastogi and Mehrotra (1991, 1995), Sharma (2003), Watt (1972).

Portulaca quadrifida Linn.

Syn.: *P. meridiana* Linn.; *P. geniculata* Royle; *Illecebrum verticillatum* Burm.

Family: Portulacaceae

Sans. and H. Names	:	*Sans.*: Laghughonika, Laghulonika;
		H.: Khatechawal, Loniya.
Distribution	:	Warmer parts of India.
Part/s Used	:	Whole Plant.
Utilization	:	Considered effective in asthma.
References	:	Ambasta (1986), Kirtikar and Basu (1984), Watt (1972).

Potentilla fruticosa Linn.

Syn.: *P. rigida* Wall.; *P. arbiuscula* Don

Family: Rosaceae

Eng. Name	:	*Eng.*: Shrubby cingue foil
Distribution	:	Temperate and Sub-alpine Himalaya from Kashmir to Sikkim: 1,700-3,000m.
Part/s Used	:	Roots.

Utilization	:	Tea prepared by boiling the roots in water till extract become reddish given to patients suffering from asthma.
References	:	Kumar and Rao (2001), Singh (1996), Kirtikar and Basu (1984), Watt (1972).

Pothos scandens Linn.
Family: Arecaceae

Distribution	:	Throughout India.
Part/s Used	:	Stem.
Utilization	:	Stems are cut and soaked with camphor for relief in asthma.
References	:	Ambasta (1986), Kirtikar and Basu (1984).

Pratia begnonifolia Lindl.
Family: Campanulaceae

Part/s Used	:	Whole Plant.
Utilization	:	Decoction of whole plant used as a remedy for asthma.
Reference	:	Kumar (2002).

Premna herbacea Roxb.
Syn.: *Pygacopremna herbacea* Moldenke; *P. pygaea* Wall.
Family: Verbenaceae

Eng., Sans. and H. Names	:	Eng.: Bhumjambu;
		Sans.: Matia jam;
		H.: Bharangi.
Distribution	:	Sub-tropical Himalaya and W. Ghats.
Part/s Used	:	Roots and Rootstock.
Utilization	:	Fresh rootstock and roots given along with ginger in asthma.
Active Constituents	:	Pygaeocins B and C and pygacone isolated from roots.
References	:	Ambasta (1986), Kirtikar and Basu (1984), Rastogi and Mehrotra (1993, 1995), Watt (1972).

Prosopis spicigera Linn.
Syn.: *P. spicata* Burm.; *Adenanthera aculcata* Roxb.; *Prospis cineraria* (L.) Druce; *Mimosa cineraria* L.
Family: Fabaceae

Sans. and H. Names	:	Sans.: Bhadra, Ishana, Ishani, Ishta;
		H.: Chhikura, Chhokara, Jhand.

Distribution	:	Punjab. Rajputana. Bundelkhand and Gujarat.
Part/s Used	:	Bark and Leaves.
Utilization	:	Decoction of leaves and bark cures asthma.
References	:	Bhatt *et al.* (2002), Kirtikar and Basu (1984), Watt (1972).

Prunus amygdalus Batsch

**Syn.: *Prunus dulcis* (Mill.) D.A. Webb.; *Amygdalus communis* L.;
A. dulis Mill.; *P. communis* (L.) Arcang**

Family: Rosaceae

Eng., Sans. and H. Names	:	*Eng.*: Almond tree;
		Sans.: Badama, Suphala, Vatada;
		H.: Badam.
Distribution	:	Cultivated in Kashmir and Punjab.
Part/s Used	:	Seeds.
Utilization	:	Emulsion good for asthma; and its juice mixed with sugar useful in cough.
Active Constituents	:	These are amygdaline, prunasin, daucosterin and sitosterol (bitter seeds) and daucosterin and sitosterol (sweet almonds).
Biological Activity	:	Antispasmodic effect confirmed.
References	:	Bhattacharjee (1998), Duke *et al.* (2002), Greenish (1999), Kirtikar and Basu (1984), Rastogi and Mehrotra (1991).

Prunus persica Batsch

**Syn.: *Amygdalus persica* Linn.; *A. collinus* Wall.;
Persica saligna Royle; *P. vulgaris* Miller.**

Family: Rosaceae

Eng. and H. Names	:	*Eng.*: Peach, Nectarine;
		H.: Aru, Shaftalu.
Distribution	:	Plains of N. India and Manipur.
Part/s Used	:	Leaves and Bark.
Utilization	:	Infusion of leaves or bark useful in cough and chronic bronchitis.
Active Constituents	:	Hentriacontane, β-sitosterol, its glucoside, ursolic acid, mandelic acid, quercetin, dotriacontane, heneicosane, tritetracontane, cyclododecyne, ethyl hexadecanoate, dotriacontan-1-ol, squalene, kaemferol and vit. E isolated from leaves. Naringenin, aromadendrin, 5, 3'-

dihydroxy-7, 4'-dimethoxyflavanone and its 5-glucoside isolated from bark.

References	:	Ambasta (1986), Kirtikar and Basu (1984), Rastogi and Mehrotra (1990, 1991, 1995), Watt (1972).

Prunus serotina Ehrh.
Family: Rosaceae

Eng. Names	:	*Eng.*: Wild black cherry.
Part/s Used	:	Bark.
Utilization	:	Bark considered useful in tuberculosis. Also, its decoction relieves cough in phthisis and bronchitis.
Active Constituents	:	Bark affords cyanogenetic glycoside prunasin and enzyme prunase.
References	:	Ambasta (1986), Bhattacharjee (1998).

Prunus virginiana Linn.
Family: Rosaceae

Eng. Name	:	*Eng.*: Common chokecherry.
Part/s Used	:	Leaves, Bark and Roots.
Utilization	:	Tea made from leaves, bark and dried roots beneficial for lung ailments.
Reference	:	Bhattacharjee (1998).

Pseudarthria viscida Wight and Arn.
Family: Fabaceae

Sans. Name	:	*Sans.*: Sanaparni.
Distribution	:	W. Peninsula.
Part/s Used	:	Roots.
Utilization	:	Roots (decoction or powder) used for asthma.
References	:	Ambasta (1986), Chopra *et al.* (1956).

Psidium guajava Linn.
[Pl. 4C]
Family: Myrtaceae

Eng., Sans. and H. Names	:	*Eng.*: Common guava;
		Sans.: Mansala, Mridue, Madhurmala;
		H.: Amrud, Safedsafari.

Distribution	:	Cultivated throughout India.
Part/s Used	:	Leaves and Flowers.
Utilization	:	Decoction of soft leaves admirable for cough and cold. Flowers used in bronchitis.
Active Constituents	:	Principle constituents are quercetin-3-α-arabino-pyranoside, sesquiguavaene, guajavarin, isoquercetin, hyperin, quercitrin, quercetin-3-ogentiobiocide, capric, lauric, myristic, palmitic, stearic and oleic acid (leaves), guavins A, C, D, ellagic acid, guaijaverin, quercetin and oleanolic acid (flowers).
References	:	Ambasta (1986), Gill *et al.* (1993), Jain *et al.* (1997), Kumar (2002), Kirtikar and Basu (1984), Rastogi and Mehrotra (1990, 1995).

Pterocarpus marsupium Roxb.

Family: Fabaceae

Eng. and H. Names	:	*Eng.*: Indian kino tree, Malabar kino tree;
		H.: Bija, Bijasal, Biya.
Distribution	:	W. Peninsula India.
Part/s Used	:	Whole Plant.
Utilization	:	Useful in asthma.
Active Constituents	:	Hydrobenzoin- Marsupol isolated.
References	:	Kirtikar and Basu (1984), Rastogi and Mehrotra (1993), Sinha *et al.* (2002).

Pterocarpus santalinoides DC.

Family: Fabaceae

Eng., Sans. and H. Names	:	*Eng.*: Red sandal wood, Ruby wood,Red sanders;
		Sans.: Arka, Bhaskarpriya, Raktasara, Ranjana;
		H.: Lalchandan, Ragatchandan.
Distribution	:	Hills of Cuddapah. S. Kurmool.
Part/s Used	:	Roots.
Utilization	:	Patients suffering from asthma given bath with water having roots boiled in it.
Active Constituents	:	Roots rich in alkaloids, saponins, resins and tannins.
References	:	Gill and Nyawuame (1994), Kirtikar and Basu (1984).

Pterospermum xylocarpum (Gaertn.) Sant. and Wagh.
Syn.: *P. heyneanum* Wall. *ex* Wight and Arn.; *Velaga xylocarpa* Gaertn.
Family: Sterculiaceae

Distribution	:	W. Peninsula.
Part/s Used	:	Bark and Leaves.
Utilization	:	Decoction of 10g bark, 5g leaves and a few black pepper given for asthma
References	:	Brahmam and Saxena (1990), Kirtikar and Basu (1984).

Punica granatum Linn.
[Pl. 5A]
Family: Punicaceae

Eng., Sans. and H. Names	:	*Eng.*: Pomegranate tree;
		Sans.: Dadima, Madhubija, Suphala;
		H.: Anar, Dharimb.
Distribution	:	Cultivated in many parts of India.
Part/s Used	:	Buds, Fruits, Seeds and Bark.
Utilization	:	Rind of roasted fruits, taken with salt for cough. Flower buds, bark and seeds useful in bronchitis.
Active Constituents	:	Sugars, vit. C, sitosterol, ursolic acid, protein, fat, mineral matter, nicotinic acid, pectin, riboflavin, thiamine, delphinidindiglycoside, asparatic, citric ellagic, gallic and malic acids, glutamine, isoquercetin, estrone and punicic acid isolated from fruit.
Biological Activity	:	Aqueous extract of roots inhibits activity of *Mycobacterium tuberculosis*.
References	:	Ambasta (1986), Anonymous (1999), Das and Agarwal (1991), Kirtikar and Basu (1984), Raju (2000), Rastogi and Mehrotra (1991, 1995).

Quercus velutina Lam.
Family: Fagaceae

Eng. Name	:	*Eng.*: Black oak.
Part/s Used	:	Bark.
Utilization	:	Decoction of bark considered effective for tuberculosis.
Reference	:	Bhattacharjee (2001).

Ranunculus scleratus Linn.

Syn.: *R. indicus* Roxb.
Family: Ranunculaceae

Eng. Names:		*Eng.*: Blister Buttercup, Marsh Crawfoot.
Distribution	:	N. India, Mt. Abu, Warm valleys of Himalaya and Bengal.
Part/s Used	:	Whole Plant.
Utilization	:	Plant extract help in relieving asthma and pneumonia.
Active Constituents	:	Active principles are anemonin and protoanemonin.
References	:	Ambasta (1986), Kirtikar and Basu (1984), Watt (1972).

Raphanus sativus Linn.

Family: Brassicaceae

Eng., Sans. and H. Names	:	*Eng.*: Radish;
		Sans.: Mulaka;
		H.: Mooli, Mula, Mura, Muri.
Distribution	:	Cultivated all over India upto 3,000m in Himalaya.
Part/s Used	:	Roots.
Utilization	:	Raw roots consumed to get relief in case of congestion or dry cough. Syrup of radishes excellent for bronchial difficulty of breathing and whooping cough.
Active Constituents	:	β-carboline derivative, trans-4- (methylthio) -3-butenyl glucosinolate isolated from roots.
References	:	Ambasta (1986), Anonymous (1999), Caius (2003), Pandey (2003), Rastogi and Mehrotra (1995).

Rheum emodi Wall. *ex* Meissn.

Syn.: *R. australe* D. Don
Family: Polygonaceae

Eng., Sans. and H. Names	:	*Eng.*: Indian Rhubarb;
		Sans.: Gandhini, Pita, Pitimulika, Revat chini.
		H.: Dohi, Revandehini.
Distribution	:	Himalayas: 3000- 4000m.
Part/s Used	:	Leaves, Roots and Rhizome.
Utilization	:	Leaves consumed as a pot herb for chronic bronchitis. Roots and rhizome used to cure asthma and bronchitis.

Active Constituents	:	Principle components are rhein, emodin and anthraquinone (roots) and 1.34 per cent oxalic acid (leaves).
References	:	Gaur *et al.* (1983), Jain (1968), Kirtikar and Basu (1984), Kaul (1997), Negi *et al.* (1999).

Rhodiola imbricata Edge.
Syn.: *Sedum imbricatum* (Edge) Walpers.; *S. rohodiola* auct. non DC.
Family: Crassulaceae

Part/s Used	:	Leaves.
Utilization	:	Decoction of leaves used as a remedy for cough and asthma.
Reference	:	Singh *et al.* (1996).

Rhododendron anthopogon D. Don
Syn.: *R. aromaticum* Wall.; *Azalea lapponica* Pall; *Osmothamnus fragrans* and *pallidus* DC.; *R. haemonium* Balf.f. and Cooper
Family: Ericaceae

Distribution	:	Alpine Himalaya: 3, 650-5,300m.
Part/s Used	:	Leaves.
Utilization	:	Decoction of leaves allays cough, cold and chronic bronchitis.
Active Constituents	:	Chief constituents of leaves are quercetin-3-O-α-L-rhamnopyranoside, kaempferol, its 4-methyl ether, 3-O-glucoside, β-sitosterol, friedelin, ursolic acid and quercetin.
References	:	Bhattacharjee (2001), Rajan *et al.* (2002), Kirtikar and Basu (1984), Rastogi and Mehrotra (1991, 1995), Watt (1972).

Rhus glabra Linn.
Family: Anacardiaceae

Part/s Used	:	Leaves.
Utilization	:	Leaves smoked for asthma.
References	:	Bhattacharjee (1998).

Rhynchanthus longiflorous Hook.f.
Family: Zingiberaceae

Part/s Used	:	Rhizome.

Utilization	:	10 ml of rhizome paste help in relieving cough.
Reference	:	Tripathi and Goel (2001).

Ricinus communis Linn.

Syn.: *Ricinus inermis* Facq.; *R. levidus* Facq.; *R. speciosus* Burm.; *R. spectabilis* Blume; *R. viridis* willd.; *Croton spinosus* Linn.

Family: Euphorbiaceae

Eng., Sans. and H. Names	:	*Eng.*: Castor oil plant, Palma cristi;
		Sans.: Amanda, Chitrabija, Ishta, Kanta, Vatari;
		H.: Arand, Erandi, Rand.
Distribution	:	Wild or cultivated throughout tropics.
Part/s Used	:	Roots and Seeds.
Utilization	:	Root and its bark used in asthma and bronchitis. Seeds also considered good in asthma.
Active Constituents	:	Seeds contain fixed oil–castrol oil (45-50 per cent), crystalline ricinine, poisonous taxalbumin ricin, stearin, tripalmitin, triricinoleine, proteins, ricinoleic acid, chymase and lysine-2-C14.
References	:	Anonymous (1999), Caius (2003), Kirtikar and Basu (1984), Sharma (2003) Singh (1999), Sinha and Sinha (2001), Rastogi and Mehrotra (1990), Watt (1972).

Rorippa nasturtium-aquaticum (L.) Hayek

Syn.: *Sisymbrium nasturtium-aquaticum* L.; *Nasturtium officinale* R.Br.

Family: Brassicaceae

Eng. Name	:	*Eng.*: Water cress.
Part/s Used	:	Whole Plant.
Utilization	:	2tsp of plant juice given for tuberculosis, pneumonia and pulmonary ailments, thrice daily until cure.
References	:	Ambasta (1986), Bhattacharjee (1998), Raju (2000).

Rorippa rubra Blackw.

Syn.: *R. gallica* Linn.

Family: Brassicaceae

Eng. Name	:	*Eng.*: French Rose.
Part/s Used	:	Petals.

Utilization	:	Rose water prepared from petals used in bronchial asthma.
Reference	:	Ambasta (1986).

Rosmarinus officinalis Linn.

Family: Lamiaceae

Eng. and H. Names	:	*Eng.*: Rosemary oil;
		H.: Rusmari.
Distribution	:	Grown in Nilgiri hills and Temperate Himalaya.
Part/s Used	:	Leaves.
Utilization	:	Dried leaves smoked to get relief from asthma.
Active Constituents	:	Chief constituents of leaves are rosemanol, carnosol, rosmadial, rosamaricine (0.33 per cent) along with another base carnosic, 4'5-dihydroxy-7-methoxyflavone and 6-methoxyluteolin glycoside; epirosamanol.
References	:	Bhattacharjee (2001), Rastogi and Mehrotra (1990, 1993, 1995), Sharma and Sood (1997).

Rotata rotundifolia (Roxb.) Kochne.

Syn.: *Ammania rotundifolia* Ham. *ex* Roxb.

Family: Lythraceae

Part/s Used	:	Leaves and Stem.
Utilization	:	Extract of stem and leaves given, twice a day, for cough.
References	:	Bennet (1987), Jha and Varma (1996).

Rumex hastatus D. Don

Family: Polygonaceae

Distribution	:	Kumaon to Kashmir: 350- 2,700m.
Part/s Used	:	Roots.
Utilization	:	Crushed roots mixed with water relieve coughing (1 tsp twice a day).
References	:	Negi *et al.* (1999), Watt (1972).

Ruta graveolens Linn.

Syn.: *R. angustifolia* Pers.; *R. chalepensis* Wall.

Family: Rutaceae

Eng., Sans. and H. Names	:	*Eng.*: Ave grace, Common rue, Garden rue, Herb of grace;

		Sans.: Pitapushpa, Sadapaha, Somalata, Vishapaha;
		H.: Pismarum, Sadab, Satari.
Distribution	:	Cultivated throughout India.
Part/s Used	:	Whole Plant.
Utilization	:	Infusion of plant generally prescribed on the west coast in acute bronchitis and pneumonia.
Active Constituents	:	These are kokusaginine, 1–methyl–2n–nonyl -4–quinolone (leaves, shoots, flowers), dictamine, ?-fagarine, skimmianine, graveolin, graveolinine, 2-3-dihydrokokusaginine, ?-fagarineaborine, aborinine and rutamine (whole plant).
Biological Activity	:	Antispasmodic activity confirmed.
References	:	Kirtikar and Basu (1984), Rastogi and Mehrotra (1990, 1991), Watt (1972).

Salacia oblonga Wall. *ex* Wight and Arn.
Family: Hippocrateaceae

Distribution	:	W. Peninsula.
Part/s Used	:	Root-bark.
Utilization	:	Root-bark (decoction or powdered) useful.
References	:	Ambasta (1986), Kirtikar and Basu (1984).

Salix nigra Marsh.
Family: Salicaceae

Eng. Name	:	*Eng.*: Black Willow.
Part/s Used	:	Bark.
Utilization	:	Infusion of bark prescribed for asthma.
Reference	:	Bhattacharjee (1998).

Salvadora persica Linn.
Syn.: *S. wightianan* Planch; *S. indica* Wight; *Cissus arborea* Forsk.
Family: Salvadoraceae

Eng. and H. Names	:	*Eng.*: Mustard tree;
		H.: Kharjal.
Distribution	:	Drier parts of India.
Part/s Used	:	Leaves.
Utilization	:	Decoction of leaves useful in asthma and cough.

Active Constituents	:	Salvadoricine isolated from leaves
References	:	Ambasta (1986), Bhatt *et al.* (2002), Kirtikar and Basu (1984), Rastogi and Mehrotra (1995), Shah (1982), Watt (1972).

Salvia coccinea Linn.
Family: Lamiaceae

Part/s Used	:	Whole Plant.
Utilization	:	Decotion of plant used for relief from tubercular bronchitis.
Active Constituents	:	Neoclerodane diterpenoid–salviacoccin, hentriaconanol, β–sitosterol, pelargonidin-3-caffeoylglucoside–5 -dimalonylglucoside and pelargonidin–3–p–coumaroyl–glucoside–5 -dimalonylglucosides isolated from flowers.
References	:	Ambasta (1986), Rastogi and Mehrotra (1991, 1995).

Salvia moorcroftiana Wall.
Family: Lamiaceae

Distribution	:	Temperate Himalaya from Kashmir to Uttarakhand upto 1,800 m.
Part/s Used	:	Roots.
Utilization	:	Fresh or dried root extract useful for cough and asthma.
Active Constituents	:	Roots yield essential oil and mucilage, 6, 7–dehydroroyleanone, diterpene- methylquinone-15-deoxyfuerstione.
References	:	Sharma (2003), Singh (1996), Rastogi and Mehrotra (1995).

Sambucus nigra Linn.
Syn.: *S. vulgaris* Lamk.
Family: Caprifoliaceae

Eng. Name	:	*Eng.*: European or Black Elder.
Distribution	:	Cultivated in India.
Part/s Used	:	Leaves, Flowers and Berries.
Utilization	:	Dried corollas and stamens used in bronchial asthma. Juice of berries useful in cold. Preparation of leaves and flowers used for cold and bronchial asthma.
Active Constituents	:	These are 0.32 per cent volatile oil, α-and β amyrins, lupeol, cycloartenol, 24–methylene cycloartanol,

cycloeuclenol, cholesterol, campesterol and stigasterol (flowers), tyrosin, sambucine cyanogenetic glucoside, sambunigrin, glucoside of laevo-phenylglycollic acid, β-sitosterol, stigasterol, campesterol, α-amyrin palmitates, ursolic, oleanolic acid and quercetin (leaves).

References   :   Ambasta (1986), Greenish (1999), Rastogi and Mehrotra (1991, 1995), Watt (1972).

Sansevieria liberica Ger.and Labr.

Family: Agavaceae

Part/s Used   :   Bark, Leaves and Roots.

Utilization   :   Decoction given to cure asthma.

Active Constituents :   Inulin and saponin isolated.

Reference   :   Gill *et al.* (1993).

Sapindus emarginatus Vahl.

Syn.: *S. trifoliatus* Linn.; *S. laurifolius* Vahl.; *S. acutus* Roxb.; *S. astergens* Roxb.

Family: Sapindaceae

Eng., Sans. and H. Names :   *Eng.*: Soap nut tree of S. India;

                             Sans.: Arishta;

                             H.: Ritha.

Distribution   :   Common in S. India.Cultivated in Bengal.

Part/s Used   :   Fruits.

Utilization   :   Extract of dried fruits given for a few days to cure asthma.

Active Constituents :   Fruits afford emariginatoside, 2 hederagenin, emarginatosides and C, kaempferol, quercetin and β-sitosterol.

References   :   Ambasta (1986), Das and Agarwal (1991), Kirtikar and Basu (1984), Rastogi and Mehrotra (1990, 1991, 1993, 1995), Singh and Pandey (1998), Watt (1972).

Sauromatum venosum (Ait) Schott.

Syn.: *Arum venosum* Ait.; *S.guttatum* (Wall.) Schott.; *S. pedatum* (Willd.) Schott.

Family: Araceae

Part/s Used   :   Corm.

Utilization	:	Powdered corm with honey useful in tuberculosis.
Reference	:	Bhogaonkar and Devarkar (2002).

Saussurea costus (Falc.) Lipsch.

Family: Asteraceae

Disrtibution	:	W. Himalaya.
Part/s Used	:	Roots.
Utilization	:	Alcoholic extract of roots effective in vagotonic type of bronchial asthma.
References	:	Kaul (1997), Shah (1982).

Saussurea lappa C.B. Clarke

Syn.: *Aucklanda costus* Falc.; *Aplotaxis lappa* Decne.

Family: Asteraceae

Eng., Sans. and H. Names	:	*Eng.*: Costus, Kuth;
		Sans.: Kakala, Kushta;
		H.: Kut, Pachak.
Distribution	:	Himalaya: 2,500-4,000m.
Part/s Used	:	Roots.
Utilization	:	Decoction of roots given for bronchial asthma and cough.
Active Constituents	:	Essential oil (1.51 per cent), saussurine (0.05 per cent), bitter resin.
Biological Activity	:	Hypotensive, spasmolytic and bronchodilatory effects found positive.
References	:	Ambasta (1986), Rajan *et al.* (2002), Bhattacharjee (1998), Caius (2003), Jain (1968), Kirtikar and Basu (1984), Kumar and Dwivedi (2000), Rastogi and Mehrotra (1990,1991), Sharma and Sood (1997), Sinha and Sinha (2001), Watt (1972).

Saussurea sorocephala (Schrenk) Sch.-Bip.

Syn.: *Aplotaxis gnaphalodes* Royle; *Saussurea gnaphalodes* (Royle *ex* DC) Sch.-Bip.

Family: Asteraceae

Part/s Used	:	Aerial Parts.
Utilization	:	Dried aerial parts (pulverized), half tsp of powder thrice a day for 5-15 days, given to cure pulmonary affections.
Reference	:	Sood *et al.* (2001).

Scindapsus officinalis Schott.
Syn.: *Pothos officinalis* Roxb.
Family: Araceae

Sans. and H. Names	:	*Sans.*: Gaja pippali;
		H.: Gajapipal.
Distribution	:	Tropical Himalaya, Bengal and Andamans.
Part/s Used	:	Fruits and Roots.
Utilization	:	Decoction of roots and fruits prescribed for asthma, cough and tuberculosis.
Active Constitutents	:	Fruit contains scindapsin A, scindapsin B, sugar and fixed oil.
References	:	Ambasta (1986), Anonymous (1999), Caius (2003), Das and Agarwal (1991), Kirtikar and Basu (1984), Rastogi and Mehrotra (1990), Saxena *et al.* (1981).

Scroparia dulcis Linn.
Family: Scrophulariaceae

Eng. Name	:	*Eng.*: Sweet Broomweed.
Distribution	:	Widespread in India.
Part/s Used	:	Leaves.
Utilization	:	Infusion of leaves given for cough and bronchitis.
Active Constituents	:	Inulin, saponin, tannin, dulcitol, scutellarein, its 7-O-methyl ether and its 7-O-β-D-glucuronide isolated from leaves.
References	:	Chopra *et al.* (1956), Karatela *et al.* (1991), Rastogi and Mehrotra (1991).

Scutellaria discolor Wall. *ex* Benth.
Family: Lamiaceae

Part/s Used	:	Whole Plant.
Utilization	:	Decoction beneficial in bronchitis.
Reference	:	Kumar (2002).

Selinum wallichianum (DC.) Raizada *ex* Saxena
Syn.: *Selinum candollii* DC.; *S. filicifolium* (Edge.) Nasir; *S. tenuifolium* Wall. *ex* Cl.; *S. tenuifolium* var. *filicifolia* (Edge.) Cl.
Family: Apiaceae

Distribution	:	Alpine pastures from Kashmir to Uttarakhand: 2,000-4,000m.

Part/s Used	:	Roots.
Utilization	:	Recommended for cough and asthma.
Active Constituents	:	Contains angelicinxanthotoxol, heraclenin, bergapten, imperatorin, heraclenol and 8-gerynyloxy-psoration.
References	:	Aswal (1996), Negi *et al.* (1993), Rastogi and Mehrotra (1991), Sharma (2003).

Semicarpus anacardium Linn.

Syn.: *S. latifolius* Pers.; *Anacardiunm latifolium* Lamk.; *A. officinarum* Gaertn.

Family: Anacardiaceae

Eng., Sans. and H. Names	:	*Eng.*: Marking Nut, Oriental Cashew Nut;
		Sans.: Agnika, Aruskara, Avhala, Bhalli;
		H.: Anala, Arushkara, Belatak, Bhela.
Distribution	:	Outer Himalaya and Warmer parts of country.
Part/s Used	:	Seeds and Fruits.
Utilization	:	Rind of the fruit cures bronchitis. Nuts used in Goa for asthma after having been steeped in buttermilk. Seed oil given to relieve cough and cold.
Active Constituents	:	Fruit kernels yield a oil comprising palmitic, stearic, linoleic, arachidic, oleic and myristic acid.
References	:	Anonymous (1999), Kirtikar and Basu (1984), Rastogi and Mehrotra (1990), Sudhakar and Rolla (1985), Tarafdar (1987), Thakur *et al.* (1989), Varma (1997), Watt (1972).

Serenoa serrulata (Michx.) Hook.

Family: Arecaceae

Eng. Name	:	*Eng.*: Saw palmetto.
Part/s Used	:	Ripe fruits.
Utilization	:	Dried ripe fruit beneficial for respiratory affections.
Reference	:	Bhattacharjee (1998).

Sesamum indicum Linn.

Syn.: *S. orientale* Linn.; *S. luteum* Retz.; *S. occidentale* Heer and Regel.

Family: Pedaliaceae

Eng., Sans. and H. Names	:	*Eng.*: Seasame;

		Sans.: Pavitra, Puraphala, Vanodbhava;
		H.: Bariktel, Til, Tilkatel.
Distribution	:	Cultivated in most parts of India.
Part/s Used	:	Seeds and Leaves.
Utilization	:	Oil from seeds useful in dry cough and asthma. Laddoos prepared by pounding its seeds with baked flowers of *Madhuca indica* given to check cough and cold. Infusion of the leaves given as a remedy for diseases of respiratory tract.
Active Constituents	:	Principle ingredients are pedalin (leaves). 3 sesaminol glucosides (VI, VII and VIII), sesaminol, sesamolinol, pinoresinol–4'o–β–D–glucosyl (1→6)–β–D glucoside and pinoresinol–4'–o–β–D–g lucosyl (1→2)- β–D-glucoside (seeds).
References	:	Kirtikar and Basu (1984), Rastogi and Mehrotra (1990, 1995), Singh and Pandey (1998), Watt (1972).

Sida acuta Burm. f.

Syn.: *S. carpinifolia* Linn.; *S. lanceolata* Roxb.; *S. stipulata* Cav.; *S. stauntoniana* DC.; *S. scroparia* Lour.

Family: Malvaceae

Sans. and H. Names	:	*Sans.*: Bala, Brihannagaba, Pata, Pila, Rajbala;
		H.: Bariara, Kareta, Kharenti.
Distribution	:	Throughout the hotter parts of India.
Part/s Used	:	Leaves and Roots.
Utilization	:	Extract of roots effective in bronchitis and pneumonia, leaves for tuberculosis.
Active Constituents	:	Fats and oil, lignin, saponin and tannins isolated from roots.
References	:	Das (1997), Gill *et al.* (1993), Kirtikar and Basu (1984), Watt (1972).

Sida cordata (Burm. f.) Bross.

Syn.: *Melochia cordata* Burm.; *Sida multicaulis* Cav.; *S. humilis* Cav.; *S. humilis* var. *varonicaefolia* (Lamk.) Masters.; *S. veronicaefolia* Lamk.

Family: Malvaceae

Part/s Used	:	Whole Plant.
Utilization	:	Decoction is said to be useful for treating asthma.

Active Constituents	:	Plant yields N-methyl Ψ-ephedrine, N-methyl ephedrine, β-phenethylamine, ephedrine, Ψ-ephedrine, vasicinol, vasicinone, vasicine, choline and betaine.
References	:	Rajwar (1983), Rastogi and Mehrotra (1993, 1995), Singh and Kumar (2000).

Sida cordifolia Linn.

Syn.: *S. herbacea* Cav.; *S. Micans* Cav.; *S. rotundifolia* Cav.

Family: Malvaceae

Sans. and H. Names	:	*Sans.*: Badiyalaka, Balini, Bhadra, Jayanti;
		H.: Bariar, Kharenti, Kungyi.
Distribution	:	Moist places throughout Tropical and Sub -tropical regions.
Part/s Used	:	Fruits.
Utilization	:	Decoction of fruits useful in bronchitis.
Active Constituents	:	Major constituents are ephedrine, mucilage, some oil and minerals.
References	:	Das and Agarwal (1991), Kitikar and Basu (1984), Sinha and Sinha (2001), Watt (1972).

Sida rhombifolia Linn.

Syn.: *S. canariensis* Willd.; *S. cempressa* Wall.; *S. rhombifolia* var. *rhomboidea* (DC.) Masters; *S. rhombifolia* var. *obovata* Wall. *ex* Masters; *S. orientalis* Cav.

Family: Malvaceae

Sans. Names	:	*Sans.*: Ahikhanda, Atibala, Barela, Devasha.
Distribution	:	Wasteplaces throughout plains of India.
Part/s Used	:	Whole Plant.
Utilization	:	10g powdered plant given, twice a day with water for 2 months, to cure pulmonary tuberculosis.
Active Constituents	:	Ephedrine is a chief constituent of leaves.
References	:	Khanna *et al.* (1996), Kirtikar and Basu (1984), Singh and Pandey (1980), Sinha and Sinha (2001), Watt (1972).

Sisymbrium irio Linn.

Family: Brassicaceae

Eng. and H. Names	:	*Eng.*: London rocket;
		H.: Khubkalan.

Distribution	:	N. India. Rajputana.
Part/s Used	:	Seeds.
Utilization	:	Decoction of seeds helps in curing asthma.
Active Constituents	:	Isorhamnetin isolated from seeds.
References	:	Ambasta (1986), Chopra *et al.* (1956), Rastogi and Mehrotra (1990).

Solanum indicum Linn.

Syn.: *S. ferox* L., *S. anguivi* Lamk.; *S. lasiocarpum* Blume; *S. hirsutum* Roxb.; *S. stramonifolium* Dunal; *S.violaceum* Facq.; *S.cuneatum* Moench.; *S.canescens* Blume; *S. heynii* Roem and Sch.; *S. pinnatifidum* Roth; *S. anguivi* Bojer.; *S. himalense* Dunal.

Family: Solanaceae

Eng., Sans. and H. Names	:	*Eng.*: Poison berry;
		Sans.: Nakuli, Sitakanta, Vrihati;
		H.: Barhanta, Barhata.
Distribution	:	Throughout warmer parts of country upto 1,500m.
Part/s Used	:	Roots and Fruits.
Utilization	:	Decoction of roots useful in cough, catarrhal affections and asthma. Fruits useful in bronchitis.
Active Constituents	:	Indioside A, carpesterol and 3 β-(p-hydroxy) benzoyloxy-22-α-hydroxy-4-α-methyl-5-α-stigast-7-en-6 isolated from fruits.
References	:	Ambasta (1986), Anonymous (1999), Bhatt *et al.* (1999), Kirtikar and Basu (1984), Kumar (2002), Rastogi and Mehrotra (1993, 1995), Sudhakar and Rolla (1985), Watt (1972).

Solanum melongena Linn.
[Pl. 5B]

Syn.: *S. incanum* Linn.; *S. insanum* Linn.; *S. insanum* Roxb.; *S. esculentum* Dunal.; *S. longum* Roxb.; *S. trongum* Lamk.; *S. torvum* var. *inerme* Dalz. and Gib.

Family: Solanaceae

Eng., Sans. and H. Names	:	*Eng.*: Brinjal, Eggplant;
		Sans.: Kantapatrica, Vartaku, Vatigama;
		H.: Baingan, Bhanta, Brinjal.
Distribution	:	Widely cultivated in India.

Part/s Used	:	Roots and Leaves.
Utilization	:	Roots considered antiasthmatic. Powdered leaves beneficial for asthma and bronchitis.
Active Constituents	:	4-ethylcatechol, trans-caffeic, hydrocaffeic, protocatechuic and chlorogenic acid isolated from leaves.
References	:	Ambasta (1986), Kirtikar and Basu (1984), Rastogi and Mehrotra (1991), Watt (1972).

Solanum myriacanthium Dunal.

Syn.: *S. khasianum* Clarke; *S. aculeatissimum* Jacq.

Family: Solanaceae

Distribution	:	North eastern India upto 2,000m.
Part/s Used	:	Fruits.
Utilization	:	Recommended for asthma, bronchitis, cough and cold.
Active Constituents	:	Solasonine, solasodine, solakhasianin, β -sitosterol, lanosterol and diosgenin and solasodine isolated from mature fruits.
References	:	Kumar (2002), Rastogi and Mehrotra (1990, 1991)

Solanum nigrum Linn.

Syn.: *S. rubrum* Mill.; *S. triangulare* Lamk.; *S. villosum* Lamk.; *S. nodiflorum* Facq.; *S. incertum* Dunal; *S. roxburghii* Dunal

Family: Solanaceae

Eng., Sans. and H. Names	:	*Eng.*: Black Nightshade, Common Nightshade;
		Sans.: Bahuphala, Katuphala, Svadupaka;
		H.: Gurkamai, Makoi.
Distribution	:	Throughout India upto 2,743m in Himalaya.
Part/s Used	:	Flowers and Fruits.
Utilization	:	Decoction of fruits alone or in combination with flowers cures cough and bronchial irritations effectively.
Active Constituents	:	Immature green fruit contains solamargine, solasonine, α-solanigrine and steroidal alkaloids–SN–O -, SN -1, SN2, SN-3, SN-4, SN-a, SN-b, SN-c, SN-d, SN-e, SN-f.
References	:	Anonymous (1999), Bhattacharjee (1998), Das and Agarwal (1991), Das *et al.* (1983), Kirtikar and Basu (1984), Rastogi and Mehrotra (1990, 1991, 1993, 1995), Sharma (2004), Swami and Gupta (1996), Watt (1972).

Solanum surattense Burn. F.

Syn.: *S. xanthocarpum* Schradt and Wenlle.; *S. jacquinii* Willd.; *S. diffusum* Roxb.; *S. mccanni* Sant.; *S. virginianum* Facq.; *S. armatum* Br.

Family: Solanaceae

Eng., Sans. and H. Names	:	*Eng.*: Yellow- Berried Nightshade;
		Sans.: Kantakari;
		H.: Kateli, Katai.
Distribution	:	Distributed throughout India in wasteplaces along roadsides.
Part/s Used	:	Roots, Whole Plant, Berry, Leaves and Bulb.
Utilization	:	Leaf infusion given in severe cough and asthma. Decoction of fruits (1tsp twice daily) taken for cough and asthma problems. Rind of fruit used for cough. Fruit and flowers mixed with ghee also useful for cough. Decoction of dried powdered plant with bark of *Adhatoda vasica* and *Piper longum* in half litre of water, beneficial in cough. Roots form a constituent of well known Ayurvedic preparation 'Dasamula', employed in cough and asthma as expectorant.
Active Constituents	:	Fruits yield carpestral, solanocarpine, solasodine, solasonine, solamargine and β–solamargine.
References	:	Ambasta (1986), Bhattacharjee (1998), Das and Agarwal (1991), Das and Misra (1988), Gupta (1997), Jain (1968), Kirtikar and Basu (1984), Mehrotra *et al.* (1995), Nagendraprasad and Abraham (1984), Rao and Neogi (1980), Shah *et al.* (1983), Singh (2002), Sinha and Sinha (2001), Watt (1972).

Solanum trilobum Linn.

Family: Solanaceae

Distribution	:	Gujarat, Deccan and Carnatic.
Part/s Used	:	Leaves and Roots.
Utilization	:	Decoction of plant useful in chronic bronchitis. Flowers and berries given in cough. Roots and leaves also recommended for bronchitis.
Active Constituents	:	Solasodine occurs in different parts of plants. Leaves contain Hignesh.
References	:	Bhattacharjee (1998), Chopra *et al.* (1956), Rastogi and Mehrotra (1995).

Solanum tuberosum Linn.

Family: Solanaceae

Eng. and H. Names	:	*Eng.*: Potato;
		H.: Alu.
Part/s Used	:	Leaves.
Utilization	:	Extract of leaves antispasmodic in cough.
Active Constituents	:	Leaves afford cycloartenol, stigasterol, β–sitosterol, campesterol, cholesterol, a mixture of 4α -monomethyl sterol; chlorogenic, neochlorogenic and isochlorogenic, 1–caffeoylquinic acid and glycosides of quercetin.
References	:	Ambasta (1986), Rastogi and Mehrotra (1990, 1991).

Solanum virginianum Linn.

Family: Solanaceae

Part/s Used	:	Whole Plant, Roots and Fruits.
Utilization	:	5g roots with equal quantity of seeds of *Piper nigrum* boiled in 250 ml water to prepare concentrate, given twice a day to cure cough. 40-50 ml decoction of whole plant with *Ocimum sanctum* given to treat cough and asthma twice a day for required period. Flowers smashed and chewed, twice daily for 2-3 days, to cure cough. Dried fruits smoked to cure cough. Extract of roots taken orally to cure cough, asthma.
References	:	Khanna and Kumar (2000), Panthi and Chaudhary (2003), Singh and Pandey (1998).

Sonchus arvensis Linn.

Syn.: *S. wightianus* DC.

Family: Asteraceae

Eng. and H. Names	:	*Eng.*: Corn sow thistle;
		H.: Sadhi, Sahadevi Bari.
Distribution	:	Throughout India; wild or cultivated.
Part/s Used	:	Roots.
Utilization	:	Roots useful in cough, bronchitis and asthma.
References	:	Ambasta (1986), Bhalla *et al.* (1996), Caius (2003), Kirtikar and Basu (1984).

Sorbus aucuparia Linn.

Syn.: *Pyrus aucuparia* Gaertn.

Family: Rosaceae

Eng. Names	:	*Eng.*: Mountain Ash, Rowan tree.
Distribution	:	W. Temperate Himalaya from Kashmir to Kumaon: 3,000- 4,300m.
Part/s Used	:	Leaves.
Utilization	:	Infusion of leaves used as pectoral in cough and bronchitis.
References	:	Ambasta (1986), Chopra *et al.* (1956).

Sphaeranthus indicus Linn.

Family: Asteraceae

Sans. and H. Names	:	*Sans.*: Avyatha, Bhukanda, Mahamundi, Shravana;
		H.: Gorakmundi, Mundi.
Distribution	:	Throughout India, ascending Himalaya upto 1,700m.
Part/s Used	:	Stem and Seeds.
Utilization	:	Decoction of seeds credited with antitubercular properties. Stem given for cough with honey.
Active Constituents	:	Sphaeranthine, β–sitosterol, stigmasterol, cadinene, β–caryophyllene, eugenol and ocimene isolated from the plant.
References	:	Ambasta (1986), Bhalla *et al.* (1996), Bhandary and Chandrashekhar (2002), Kirtikar and Basu (1984), Sharma (2004), Sinha and Sinha (2001).

Stellaria pubera Michx.

Family: Caryophyllaceae

Eng. Name	:	*Eng.*: Giant chick weed.
Part/s Used	:	Whole Plant.
Utilization	:	Infusion of plant helps in relieving lungs congestion and also effective in tuberculosis.
Reference	:	Bhattacharjee (1998).

Stephania glabra Miers.

Syn.:*S. rotunda* Hook.f and Thoms;
***Cissampelos glabra* Roxb.; *S delavayi* auct. non Diels.**

Family: Menispermaceae

Distribution	:	Himalayas from Shimla to Sikkim, Khasia Hills and Assam.
Part/s Used	:	Roots and Tubers.
Utilization	:	Roots useful in asthma and pulmonary tuberculosis. Powdered tubers mixed with honey given to relieve asthmatic attacks and also recommended for tuberculosis.
Active Constituents	:	Active alkaloids are palmatine, dehydrocorydalmine, palmatrubine, stepharanine, tetrahydropalmatine, corydalmine, stephollidine and cycleamine.
References	:	Ambasta (1986), Bennet (1987), Chopra *et al.* (1956), Kirtikar and Basu (1984), Kumar (2002), Rao and Jamir (1982), Rastogi and Mehrotra (1991, 1993, 1995).

Stercularia rubiginosa Vent. var. *glabrescens* King.

Family: Sterculiaceae

Parts used	:	Leaves.
Utilization	:	Nicobares use decoction of leaves in morning and at bed time for asthma and cough.
Reference	:	Sinha (1996).

Stereospermum chelonoides (L.f.) DC.

Syn.: *S. suaveolens* (Roxb.) DC.; *Bignonia chelonoides* Linn.;
***Heterophraga chelonoides* Dalz and Gibs.; *B. suaveolens* Roxb.**

Family: Bignoniaceae

Sans. and H. Names	:	*Sans.*: Ambuvasi, Kalavrinta, Kuberakshi;
		H.: Padal, Paral, Purula.
Distribution	:	Throughout India in drier localities; often planted.
Part/s Used	:	Roots.
Utilization	:	Roots allay asthma.
References	:	Kirtikar and Basu (1984), Malhotra and Moorthy (1973), Watt (1972).

Stereospermum personatum (Hassk.) DC.

Syn.: *S. chelonoides* auct; *S. tetragonum* A. DC.

Family: Bignoniaceae

Eng. and H. Names	:	*Eng.*: Trumpet flower, Yellow snake tree;
		H.: Pader, Padri, Parrel.
Part/s Used	:	Roots.
Utilization	:	Decoction of roots used to relieve asthma and cough.
References	:	Ambasta (1986), Bennet (1987).

Strobilanthes humilis Gamble.

Family: Acanthaceae

Part/s Used	:	Whole Plant.
Utilization	:	Powdered 15- 20g plant cooked with 8g garlic in 20ml water, prescribed with tea or any drink 3 times per day for 7 days for asthma.
Reference	:	Rajendran and Mehrotra (1996).

Styrax benzoin Dryanl.

Family: Styraceae

Eng. Name	:	*Eng.*: Benzoin tree.
Part/s Used	:	Resin.
Utilization	:	Vapours of resins added to boiling water are inhaled in case of bronchitis.
Active Constituents	:	Fresh resin contains 79 per cent crystalline coniferyl benzoate and 12 per cent benzoic acid.
References	:	Bhattacharjee (1998), Chopra *et al.* (1956).

Strychnos ignatii Berg.

Syn.: *Ignatia amara* Linn.; *S. maingayi* Cl. var *fructuosa* Cl.; *S. pseudotieute* Hill.

Family: Loganiaceae

H. Name	:	*H.*: Pipeta.
Distribution	:	Cultivated in gardens.
Part/s Used	:	Seeds.
Utilization	:	1-2 grains eaten to cure asthma.

Active Constituents	:	Seeds contain 1.5 per cent strychnine and 2.5 per cent brucine.
References	:	Das and Aggarwal (1991), Greenish (1999), Watt (1972).

Strychnos nux-vomica Linn.

Syn.: *S. lucida* Wall. *S. colubrina* Wight

Family: Loganiaceae

Eng., Sans. and H. Names	:	*Eng.*: Crow fig, Nux–vomica tree, Poison Nut;
		Sans.: Kakanda, Karaskara, Vartula;
		H.: Kajra, Nirmal.
Distribution	:	Throughout India upto an altitude of 1,300m.
Part/s Used	:	Stem-bark.
Utilization	:	2-3 tsp decoction of crushed stem-bark crushed with black pepper (twice a day for 45 days) considered effective for asthma by Khonds and Porjas.
Active Constituents	:	Bark contains strychnine, brucine, α and β colubrine and pseudo-strychnine.
References	:	Chopra *et al.* (1956), Kirtikar and Basu (1984) Rao and Henry (1996), Watt (1972).

Symplocarpus foetidus (L.) Salisb.

Family: Araceae

Part/s Used	:	Roots.
Utilization	:	Dried powdered roots given for cure of asthma and chest ailments.
Reference	:	Bhattacharjee (1998).

Syzygium aromaticum Merr. and Perry
[Pl. 6A]

Syn.: *Caryophyllus aromaticus* Linn.; *Eugenia caryophyllata* Thunb.; *E. aromatica* Kuntze

Family: Myrtaceae

Eng. and H. Names	:	*Eng.*: Clove tree;
		H.: Laung.
Distribution	:	Cultivated in S. India.
Utilization	:	5-10ml decoction of 2/3 cloves and 1 tsp dried rhizome of *Zingiber officinale*, once or twice daily, for cough and cold.

Active Constituents	:	Caryophyllene, eugenol and naphthaline isolated form clove buds.
References	:	Ambasta (1986), Chopra *et al.* (1956), Rastogi and Mehrotra (1991, 1993, 1995), Sen and Batra (1999).

Syzygium cumini (L.) Skeels.
[Pl. 7A]

Syn.: *Eugenia jambolana* Lam.; *E. cumini* Druce.; *Myrtus cumini* L.

Family: Myrtaceae

Eng. and H. Names	:	*Eng.*: Black plum, Jaman, Jambolan;
		H.: Jamun.
Distribution	:	M.P. Moist deciduous forest.
Part/s Used	:	Stem-bark.
Utilization	:	2tsp extract of stem-bark considered useful for cough (twice a day for 5 days) by Bagatas and Porjas.
Active Constituents	:	Betulinic acid, friedelin, friedelinol, kaemferol, quercetin and sitosterol isolated from stem-bark.
References	:	Ambasta (1986), Anonymous (1999), Bhattacharjee (1998), Jain (1968), Rao and Henry (1996), Rastogi and Mehrotra (1991).

Syzygium jambos (Lam.) Alston.

Syn.: *Eugenia jambos* L.; *Jambosa vulgaris* DC.

Family: Myrtaceae

Distribution	:	Widely in different parts of India.
Part/s Used	:	Bark.
Utilization	:	Bark given for cure of bronchitis.
Active Constituents	:	Bark affords alkaloid jambosine, tannin and oleoresin.
References	:	Bhattacharjee (1998), Chopra *et al.* (1956).

Tacca leontopetaloides (L.) O. Kuntze

Family: Taccaceae

Part/s Used	:	Stem-bark.
Utilization	:	3tsp of paste formed by mixing powdered 10g dried stem- bark with 50 ml water given, thrice a day for a week, to treat asthma and cough.
Active Constituents	:	β-sitosterol, ceryl alcohol and taccalin isolated from stem-bark.

References	:	Kumar and Pullaiah (1998), Rastogi and Mehrotra (1990).

Tagetes erecta Linn.

Family: Asteraceae

Eng., Sans. and H. Names	:	*Eng.*: African marigold, French marigold;
		Sans.: Ganduga, Sthulapushpa, Zandu;
		H.: Genda, Kalaga, Makhmali.
Distribution	:	Extensively cultivated as garden plants.
Part/s Used	:	Herb.
Utilization	:	Infusion of herb used against cold and bronchitis.
Active Constituents	:	Fresh flowering plant yields essential oil, Tagetes oil and β-sitosterol.
Biological Activity	:	Bronchodilatory and spasmolytic activities positive.
References	:	Ambasta (1986), Kirtikar and Basu (1984), Rastogi and Mehrotra (1990, 1993, 1995), Watt (1972).

Tamarindus indica Linn.

Syn.: *T. occidentalis* Gaertn.; *T. officinalis* Hook.

Family: Fabaceae

Eng., Sans. and H. Names	:	*Eng.*: Tamarind tree;
		Sans.: Amlika, Sukta;
		H.: Amli, Amlica.
Distribution	:	Cultivated throughout India.
Part/s Used	:	Fruits.
Utilization	:	Pulp of fruit used to treat bronchial asthma.
Active Constituents	:	A fruit pulp yields tamarindienal.
References	:	Jain and Tarafdar (1970), Kirtikar and Basu (1984), Rajan *et al.* (1999), Rastogi and Mehrotra (1995), Watt (1972).

Taraxacum officinale Weber

Family: Asteraceae

H. Names	:	*H.*: Barau, Kanphul.
Distribution	:	Throughout Himalayas: 3,000-6,000m.
Part/s Used	:	Flowers.
Utilization	:	Decoction of flowers (5-10 ml, twice a day) prescribed with honey for treatment of cough.

Active Constituents	:	Principle constituents are umbelliferone, esculetin, p–coumaric, caffeic, ferulic, p-hydroxbenzoic, protocatechuic, vanillic, β-resorcyclic, syringic and p-hydroxyphenylacetic acid.
References	:	Arya *et al.* (1999), Greenish (1999), Rastogi and Mehrotra (1990, 1995), Thakur *et al.* (1989).

Taxodium macronatum Tenore

Syn.: *T.distichum* var. *mucronatum* Henry

Family: Taxodiaceae

Eng. Names	:	*Eng.*: Mexican marsh Cyprus.
Distribution	:	Nilgiris, Kerala and Dehradun.
Part/s Used	:	Bark, Leaves and Roots.
Utilization	:	Useful in bronchial ailments.
Active Constituents	:	Hinokiflavone, isocryptomerin, cryptomerin A and cryptomerin B isolated.
References	:	Ambasta (1986), Rastogi and Mehrotra (1991).

Taxus baccata Linn.

Syn.: *T. nucifera* Wall.; *T. nepalensis* Facq.; *T. contorta* Griff.; *T. orientalis* Bertholini

Family: Taxaceae

Eng., Sans. and H. Names	:	*Eng.*: Common yew;
		Sans.: Manduparni;
		H.: Thuno, Birmi.
Distribution	:	Temperate Himalayas: 2,000-3,650m.
Part/s Used	:	Leaves, Fruits and Plant Extract.
Utilization	:	Decoction of leaves used as herbal tea for treating cough and cold. Jelly prepared from the fruits given for chronic bronchitis. Plant sap purified by adding cow's urine and lime water considered useful for asthma.
Active Constituents	:	Principle constituents are taxine, taxane derivatives–19 -hydroxybaccatin III, 10–deacetylcephalomannine, 10–deacetyltaxol, taxagifin, amentoflavone, mono–and di–O methylmentoflavones, sciadopitysin, ginkgetin, sequoiaflavone (leaves), rhodoxanthin and eschscholtzxanthone (fruits).
References	:	Ambasta (1986), Bhattacharjee (1998), Kaul (1997), Kiritkar and Basu (1984), Kumar and Rao (2001), Lal *et al.* (1996), Pandey and Pandey (1999), Purohit (2000), Rastogi and Mehrotra (1990, 1993, 1995), Watt (1972).

Tectaria macrodonta (Fec.) C.Chr.

Family: Aspidiaceae

Part/s Used	:	Fronds.
Utilization	:	Powdered leaves and honey or its decoction useful in asthma and bronchitis.
References	:	Bhattacharjee (2001), Kothari and Londhe (1999), Sharma and Vyas (1985), Singh and Pandey (1998).

Tectona grandis Linn.f.

Family: Verbenaceae

Eng., Sans. and H. Names	:	*Eng.*: Indian oak, Ship Tree, Teak;
		Sans.: Anila, Arna, Gandhasara, Sakila;
		H.: Sagon, Sagwan, Sagun.
Distribution	:	Konkan, W. Ghats of Bombay, Madras and C. India.
Part/s Used	:	Bark and Flowers.
Utilization	:	Considered useful in bronchitis.
Active Constituents	:	Betulin, betulin aldehyde, betulinis acid and lupeol isolated from stem-bark.
References	:	Ambasta (1986), Kirtikar and Basu (1984), Rastogi and Mehrotra (1995).

Tephrosia densifolia Guill. and Pers.

Syn.: *T.vogelii* Hook.

Family: Fabaceae

Eng. Name	:	*Eng.*: Fish poisonban.
Part/s Used	:	Leaves.
Utilization	:	Decoction of leaves allays cough.
Reference	:	Gill and Nyawuame (1994).

Tephrosia purpurea Pers.

Syn.: *T. hamiltonii* Drumm. *ex* Gamble.; *T. stricta, taalorii, wallichii, lobata, tinctoria, galegoides* and *lanceolaty* Grah.; *Galega purpurea* Linn.; *G. tinctoria* Lamk.; *G. lanceaefolia* Roxb.; *G. colonila* and *sericea* Ham.; *Cracca purpurea* Linn.; *Indigofera flexuosa* Grah.

Family: Fabaceae

Eng., Sans. and H. Names	:	*Eng.*: Wild indigo;
		Sans.: Kalika, Kalashaka, Sharapunkha;
		H.: Dhamasia, Sarphonka.

Distribution	:	Edges of scrub jungles, crop fields, water courses of Sub-montane Himalaya and adjacent plains.
Part/s Used	:	Whole Plant, Roots and Root-bark.
Utilization	:	Pulverized roots and dried root-bark smoked for relief from asthma and cough. Decoction of roots given for bronchitis, asthma. Root paste with ginger and honey recommended for cough. Decoction of dried plant given for cough.
Active Constituents	:	Coumestone- 2–methoxy–3, 9 dihydroxycoumestone, pongamol, flemichapparins B and C, rutin, methylkaranjic acid, spinasterol and ursolic acid. O - methylpongamol, lanceolatin A, lanceolatin B, (+) purpurin and maackiain, a β–hydroxychalcone - Purpurenone; isolonchocarpin, pongamol, karanjin, kanjone and β-sitosterol isolated from roots.
References	:	Ambasta (1986), Gill and Nyawuame (1994), Kirtikar and Basu (1984), Kumar and Pullaiah (1999), Rastogi and Mehrotra (1990, 1993, 1995), Raju (2000), Sharma (2003), Watt (1972).

Terminalia arjuna Wight and Arn.

Syn.: *T. berryi* W. and A.; *T. glabra* W. and A.; *T. ovalifolia* Rottl.; *Pentaptera arjuna* Roxb. *ex* DC.; *P. glabra* Roxb.

Family: Combretaceae

Eng., Sans. and H. Names	:	*Eng.*: White murdah;
		Sans.: Arjuna, Dhavala;
		H.: Anjan, Arjuna, Sadura, Savimadat.
Distribution	:	Sub-Himalayan tract and Chota Nagpur.
Part/s Used	:	Bark.
Utilization	:	Decoction of bark along with cow's milk, 1tsp twice a day, given for one month for curing asthma. Porridge containing powdered bark (3.88 g) of *T. arjuna* exposed to full moon during night is said to be useful for patients suffering from asthma, tuberculosis and bronchosis.
Active Constituents	:	Important constituents of stem-bark are oxalic acid, tannins, arjunic, arjugenin acid, arjunglucoside I and II, hentriacontane, arachidic and myrsityl stereate, arjunolone, baicalein and β-sitosterol.
References	:	Ambasta (1986), Duke *et al.* (2002), Hembrom (1991), Kirtikar and Basu (1984), Rastogi and Mehrotra (1991, 1993), Singh and Prakash (1996), Thakur *et al.* (1989), Watt (1972).

Terminalia bellirica Roxb.

Syn.: *T. bellirica* var. *laurinoides* Cl.; *Myrobalanus bellirica* Gaertn.

Family: Combretaceae

Eng., Sans. and H. Names	:	*Eng.*: Bedda nuts, Belleric myrobalan;
		Sans.: Aksha, Bahira, Bahuvirya, Harya;
		H.: Bharla, Buhura.
Distribution	:	Upto 1,000m all over India except dry and marshy areas.
Part/s Used	:	Fruits and Seeds.
Utilization	:	Powdered seeds and fruits effective for treating cough, bronchitis and asthma.
Active Constituents	:	β-sitosterol, gallic acid, ellagic acid, ethyl gallate, galloyl glucose, chebulagic acid, mannitol, bellericanin isolated from fruits.
References	:	Ambasta (1986), Bhattacharjee (1998), Kirtikar and Basu (1984), Kothari and Moorthy (1996), Rastogi and Mehrotra (1991, 1995), Shah *et al.* (1983, 1981), Singh and Pandey (1998), Sinha and Sinha (2001), Thakur *et al.* (1989).

Terminalia chebula Retz.

Syn.: *T.reticulata* Roth.; *T.aruta* Ham.; *Myrobalanus chebula* Gaertn.; *Embryogonia arborea* Teys and Binn.

Family: Combretaceae

Eng., Sans. and H. Names	:	*Eng.*: Black myrobalan, Chebulic myrobalan;
		Sans.: Abhaya, Girija, Nandini, Pathya;
		H.: Harara, Harra.
Distribution	:	Sub-Himalayan tract upto 1,000m in deciduous forests.
Part/s Used	:	Fruits.
Utilization	:	Rind of a roasted fruit chewed for instant relief from cough. Powdered fruits smoked in a pipe to provide relief in a fit of asthma.
Active Constituents	:	Chief constituents are chebulin, glycoside similar to sennoside, terchebin, gallic acid (1.21 per cent), terchebulin, punicalagin, fatty acids, reducing sugars and crude fibres.
References	:	Ambasta (1986), Kirtikar and Basu (1984), Kumar and Jain (1998), Mehrotra *et al.* (1995), Panthi and Chaudhary (2003), Rastogi and Mehrotra (1990, 1995), Sharma (2003), Singh and Singh (1990), Thakur *et al.* (1989), Watt (1972).

Thevetia peruviana (Pers.) Merr.

Syn.: *T. neriifolia* Juss.

Family: Malvaceae

Eng., Sans. and H. Names	:	*Eng.*: Lucky nut tree;
		Sans.: Ashvaghna, Haripriya;
		H.: Kulkephul, Pilekaner.
Distribution	:	Cultivated throughout India.
Part/s Used	:	Fruits.
Utilization	:	Recommended for bronchitis.
Active Constituents	:	Epiperuviol acetate, herperitin–7–glucoside, α–and β–amyrin, kaemferol and quercetin isolated from fruits.
References	:	Ambasta (1986), Das and Agarwal (1991), Rastogi and Mehrotra (1991, 1995).

Thymus vulgaris Linn.

Family: Lamiaceae

Eng. Names	:	*Eng.*: Common thyme, Garden thyme.
Distribution	:	Grown in N. India upto 900m.
Part/s Used	:	Whole Plant, Leaves, Oil from leaves and Flowers.
Utilization	:	Decoction of dried leaves mixed with ginger admirable for cough and bronchitis, and that of plant for cough and cold. Oil extracted from leaves and flowers excellent for the treatment of whooping cough and bronchitis.
Active Constituents	:	Main constituents are thymol, carvacrol (43 per cent), p-cymol (41 per cent), α-pinene, β-pinene, camphene, (-) limonene, (+) limonene, fenchone, citronellal, terpineol, borneol, citral, anethole and ethyl eugenol. p-cymene, borneol and carvacrol (volatile oil), luteolin-7-β-glucoside, luteolin-diglucoside, caffeic, ursolic, oleanolic acid, trans-4-thujanol and 4-terpineol (leaves).
References	:	Ambasta (1986), Kapur *et al.* (1996), Rastogi and Mehrotra (1990, 1993, 1995), Sharma (2003).

Tinospora cordifolia (Willd.) Hook. f. and Thoms.

Family: Menispermaceae

Eng., Sans. and H. Names	:	*Eng.*: Gulancha tinospora;
		Sans.: Amrita, Amrita lata, Amritavallil;
		H.: Amrita, Guloh.

Distribution	:	Throughout the Tropical India.
Part/s Used	:	Leaves, Stem, Fruits and Seeds.
Utilization	:	Leaves grounded with lime juice taken with honey, 3 times a day for 2 days for cure of asthma. Juice of leaves and stem considered useful in asthma. Fruit considered useful in pneumonia and flu; and seeds for curing pulmonary tuberculosis.
Active Constituents	:	Principle components are octacosanol, β-sitosterol (leaves), lignans, highly oxygenated diterpenoid, giloin, giloinin, furanolactone, tinosporidine and β-sitosterol (stem).
References	:	Ambasta (1986), Bhandary *et al.* (1996), Caius (2003), Kirtikar and Basu (1984), Kumar (2002), Mehrotra *et al.* (1995), Rastogi and Mehrotra (1991, 1995), Roy and Chaturvedi (1986), Shah (1984), Sinha and Sinha (2001), Thakur *et al.* (1989).

Tinospora sinensis Merr.

Syn.: *Campylus sinensis* Lour.; *T. tomentosa* (Colebr.) Hook. and Thoms.; *T. malabarica* (Lamk.) Hook. and Thoms.

Family: Menispermaceae

Part/s Used	:	Roots and Stem.
Utilization	:	Extract of stem and roots taken with honey affords relief in bronchitis.
Active Constituents	:	Kokasaginine, malabarolide, 10α -hydroxycolumbin, menispermacide, tinpsinen, magnoflorine, quercetin–3–O–glucoside, kaemferol and its 3-O-glucoside isolated from stem.
References	:	Rastogi and Mehrotra (1995), Saini (1996).

Toona ciliata Roem.

Family: Meliaceae

Part/s Used	:	Fruits.
Utilization	:	Powdered roasted fruits (1tsp, 3 times a day) given for treatment of cough.
Active Constituents	:	2 meliacin butenolides–21 -hydroxycederlonelide and 23 -hydroxycederlonelide, cedrelone, 23 -hydroxyto-onacilide, β–amyrin, α–amyrin, siderin, scopoletin, iosfraxidin and sitosterol isolated from fruits.
References	:	Singh and Kumar (2000), Rastogi and Mehrotra (1995).

Trachyspermum ammi (L.) Sprague
[Pl. 6B]

Syn.: *T. copticum* Link; *Carum copticum* Hiern.; *Ammi copticum* L.;
Carum ajowan Ined.; *Sison ammi; C. copticum* (L) Cl.

Family: Apiaceae

Eng., Sans. and H. Names	:	*Eng.*: Ajowan, Carum;
		Sans.: Ajmola;
		H.: Ajwain.
Part/s Used	:	Seeds.
Utilization	:	10-20g seed powder taken with hot water for relief from cough and cold. 'Ajwan oil' used as expectorant in bronchial pneumonia and other respiratory ailments.
Active Constituents	:	Essential oil, Ajowan oil mainly consists of 35-60 per cent thymol and carvacrol,fatty acids like palmitic (5.3 per cent), petroselenic (48.1 per cent), oleic (23.9 per cent) and linoleic acid (20 per cent).
References	:	Ambasta (1986), Duke *et al.* (2002), Kirtikar and Basu (1984), Panthi and Chaudhary (2003), Raju (2000), Rastogi and Mehrotra (1990, 1995), Thakur *et al.* (1989).

Tragia cannabina Linn.f.

Syn.:*T. involucrata* Linn. var. *cannabina* Muell.-Arg.;
T. plukenetii A.R. Smith; *Croton hastatus* L.; *C. urens* L.

Family: Burseraceae

Distribution	:	Dry places throughout India from Punjab and lower Himalaya of Kumaon eastwards to Assam.
Part/s Used	:	Roots.
Utilization	:	Decoction of roots used in bronchial troubles.
References	:	Ambasta (1986), Watt (1972).

Tragopogon pratensis Linn.

Family: Liliaceae

Eng. Names	:	*Eng.*: Bucks-Beard, Meadow Salsify.
Distribution	:	W. Himalaya:4,000-4,750m.
Part/s Used	:	Ligulate florets.
Utilization	:	Tinctures prepared from florets employed for respiratory diseases.
References	:	Ambasta (1986), Chopra *et al.* (1956).

Trema guineensis Fiael.

Family: Ulmaceae.

Part/s Used	:	Roots and Stem-bark.
Utilization	:	Considered useful for treating cough, bronchitis
Active Constituents	:	Inulin, lignin, saponin and tannin isolated.
Reference	:	Gill *et al.* (1993).

Trema orientalis L.

Syn.: *T. amboinenis* auct. Non Blume; *Celtis orientalis* Linn.; *Sponia wightii* Planch.; *S. orientalis* Planch

Family: Ulmaceae

Eng. and Sans. Names	:	*Eng.*: Charcoal tree, Indian nettle tree;
		Sans.: Jivani, Tivanti.
Distribution	:	Throughout India.
Part/s Used	:	Leaves.
Utilization	:	Decoction of leaves useful for cough.
Active Constituents	:	Alkaloids, lignin, inulin and tannin isolated.
References	:	Ambasta (1980), Karatela *et al.* (1991), Kirtikar and Basu (1984), Watt (1972).

Trianthema portulacastrum Linn.

Syn.: *T. monogyna* Linn.

Family: Aizoaceae

Sans. and H. Names	:	*Sans.*: Shvetapunarnava;
		H.: Lala-Sabuni, Santhi.
Distribution	:	Throughout India.
Part/s Used	:	Whole Plant and Roots.
Utilization	:	Roots used for curing asthma. 1tsp powdered plant with a pinch of fruit powder of *Piper longum* given either with honey or hot water to treat pneumonia.
Active Constituents	:	3, 4-Dimethoxycinnamic acid and β-cyanin isolated.
References	:	Ambasta (1986), Chopra *et al.* (1956), Khanna and Kumar (2000), Rastogi and Mehrotra (1993).

Tribulus terrestis Linn.

Syn.: *T. lanuginosus* Linn.

Family: Zygophyllaceae

Eng., Sans. and H. Names	:	*Eng.*: Calothrops, Calotrops root;

Sans.: Bhakshataka, Gokhura, Goksuraka;

H.: Gokhru, Gokshri.

Distribution	:	Hot and dry regions in India; upto 1,000 m in Kashmir.
Part/s Used	:	Roots, Leaves and Fruits.
Utilization	:	Powdered fruits (1 tsp, 3 times daily) prescribed for cough and asthma. Root infusion beneficial for asthma and bronchitis.
Active Constituents	:	These are campesterol, β-sitosterol, stigmasterol, diosgenin, neotogogenin, tigogenin, hecogenin and gitogenin (roots), kaempferol, kaemferol-3-glucoside, kaemferol-3-rutinoside and tribuloside (fruits and leaves).
References	:	Anonymous (1999), Arshad *et al.* (1997), Kirtikar and Basu (1984), Kothari and Rao (1999), Mehrotra *et al.* (1995), Rajan *et al.* (2002), Rastogi and Mehrotra (1990, 1993), Singh (2002), Sinha and Sinha (2001), Thakur *et al.* (1989), Watt (1972).

Trigonella foenum graecum Linn.
[Pl. 7B, 7C]
Family: Fabaceae

Eng., Sans. and H. Names	:	*Eng.*: Fenugreek, Greek Hayes;
		Sans.: Bahuparini, Bahupatrika, Methika;
		H.: Mathi, Methi.
Distribution	:	Cultivated in many parts of India.
Part/s Used	:	Seeds.
Utilization	:	Boiled or roasted seeds useful in chronic cough.
Active Constituents	:	Trigonelline, choline coumarine, quercetin, luteolin, diosgenin, gitogenin, glycoside–furost–5–en–3β, 22, 26–triol with glucose, rhamnose and xylose, titogenin, neotigogenin, 4-hydroxyisoleucine, vitexin, vicenin-1, vicenin-2, yamogenin tetrosides B and C isolated from seeds.
References	:	Ambasta (1986), Anonymous (1999), Das and Agarwal (1991), Gill and Nyawuavme (1994), Greenish (2003), Kirtikar and Basu (1984), Rastogi and Mehrotra (1990, 1991,1993).

Tussilago farfara Linn.

Family: Asteraceae

Eng. and H. Names	:	*Eng.*: Bull foot, Colt herb, Common colt's foot;
		H.: Watapana, Watpat.
Distribution	:	W. Himalaya from Kashmir to Kumaon: 2,000-3,700m.
Part/s Used	:	Leaves.
Utilization	:	Leaves smoked like tobacco as a remedy for asthma. Infusion of leaves efficacious in bronchitis.
Active Constituents	:	Leaves contain tannin, essential oil, potassium nitrate, senkirkine (0.01 per cent), p–hydroxybenzoic acid, caffeic and caffeoyltartaric acid.
References	:	Ambasta (1986), Bhattacharjee (1998), Caius (2003), Kirtikar and Basu (1984), Rastogi and Mehrotra (1993).

Tylophora indica Merr.

Syn.: *T. asthmatica* Wight and Arn.; *T. pubescens* Wall.; *T. vomitoria* Voight.; *Asclepias asthmatica* Willd.; *A. tunicata* Hort; *A. vomitoria* Koen.; *Ceynanchum vomitorium* Thwaites.; *C. ipecaeuanha* Willd.; *C. indicum* Burm.

Family: Asclepiadaceae

Eng. and H. Names	:	*Eng.*: Emetic swallow wort;
		H.: Antamul, Jangli Pikvam;
Distribution	:	Plains of India.
Part/s Used	:	Leaves, Roots and Root-bark.
Utilization	:	Decoction of leaves and infusion of root bark (1:10) used for asthma and bronchitis. Grounded leaves with onion and black pepper, taken orally to cure asthma. Dried roots considered beneficial in treating asthma, bronchitis.
Active Constituents	:	Active principles are tylophornidine (roots), (+) septicine, (+) isotylocrebrine, α–amyrin, tylophorine, kaempferol and quercetin (leaves).
References	:	Ambasta (1986), Bhattacharjee (1998), Jain (1968), Kirtikar and Basu (1984), Rastogi and Mehrotra (1990, 1991, 1993), Rosakutty *et al.* (1999), Sharma (2003), Sharma and Sood (1997), Sinha and Sinha (2001), Sudhakar and Rolla (1985).

Urginea indica (Roxb.) Kunth.

Syn.: *Scilla indica* Roxb.; *U. senegalensis* Kunth.; *S. cundria* Ham.; *U.denudata* Ham.; *Drimia indica* (Roxb.) Jessop.

Family: Liliaceae

Eng., Sans. and H. Names	:	*Eng.*: Indian squill;
		Sans.: Kolakanda, Putalu, Suputa, Vanapandan;
		H.: Jangli kanda, Jangli piyaz, Kanda.
Distribution	:	S. and E. India and N. W. Himalaya upto 2,000m.
Part/s Used	:	Corm.
Utilization	:	1-3 powdered corms used in asthma and tuberculosis.Also, its syrup given as an expectorant in chronic bronchitis.
Active Constituents	:	Bufadienolides, scillarens A and B, hentriacontanol, sitosterol and octacosanoic acid isolated from corms.
References	:	Ambasta (1986), Bhattacharjee (2001), Bhatt *et al.* (1999), Jadeja (1999), Jain (1968), Kirtikar and Basu (1984), Sharma (2003), Singh and Pandey (1998), Sinha and Sinha (2001), Thakur *et al.* (1989), Watt (1972).

Usnea sikkimensis Biswas

Family: Usneaceae

Part/s Used	:	Thallus.
Utilization	:	Decoction useful in asthma and lung troubles.
Reference	:	Ambasta (1986).

Vanda spathulata Spreng.

Family: Orchidaceae

Distribution	:	W. Peninsula from Malabar to Travancore.
Part/s Used	:	Flowers.
Utilization	:	Powdered dried flowers given for cure of asthma.
References	:	Ambasta (1986), Caius (2003), Kirtikar and Basu (1984).

Vanda tessellata Hold. *ex* Lound

Syn.: *V. roxburghii* R. Br.; *Cymbidium tesselloides* Roxb.; *C. tessellatum* Swartz.; *Epidendrum tessellatum* Roxb.; *Aerides tessellatum* Wight.

Family: Orchidaceae

Sans. and H. Names	:	*Sans.*: Atisara, Elaparni, Muktarasa, Rasya, Vandaka;
		H.: Banda, Nai, Persara, Rasna.

Distribution	:	Bengal. Jharkhand. Bihar. M.P. Chhatisgarh. W. Peninsula. Travancore.
Part/s Used	:	Roots.
Utilization	:	Roots considered useful in bronchitis.
Active Constituents	:	β- and γ–sitosterols and a long chain of aliphatic compounds isolated.
Biological Activity	:	Antitubercular property positive.
References	:	Ambasta (1986), Kirtikar and Basu (1984), Rastogi and Mehrotra (1990), Sharma (2003), Watt (1972).

Vateria indica Linn.

Syn.: *V. malabarica* Blume.

Family: Dipterocarpaceae

Eng., Sans. and H. Names	:	*Eng.*: Indian copal tree, Puiney varnish tree;
		Sans.: Ajakarna, Dhupa, Kundura;
		H.: Kahruba, Safed dammar, Sandras.
Distribution	:	W. Peninsula from Kanara to Travancore ascending to 1,350m.
Part/s Used	:	Resin.
Utilization	:	Considered useful for chronic bronchitis and throat troubles.
References	:	Ambasta (1986), Kirtikar and Basu (1984), Watt (1972).

Verbascum thapsus Linn.

Syn.: *V. indicum* Wall.

Family: Scrophulariaceae

Eng. and H. Names	:	*Eng.*: Adams flannel, Common Mullein;
		H.: Ban, Gidar, Phulla, Tamaku.
Distribution	:	Temperate Himalaya: 2,000-3,650m.
Part/s Used	:	Leaves and Roots.
Utilization	:	Dried leaves smoked in asthma, bronchitis and spasmodic cough. Roots boiled and its decoction used to treat cough. Also, its decoction used for curing asthma and other pulmonary diseases.
Active Constituents	:	Saponin, α-carotene, flavonoid–7, 4- dihydroxyflavone–4'–rhamnoside isolated from leaves.

References	:	Ambasta (1986), Bhattacharjee (1998), Chandrasekar and Srivastava (2003), Kirtikar and Basu (1984), Rajan *et al.* (2002), Rastogi and Mehrotra (1995), Sharma (2003), Sharma *et al.* (1979), Sood *et al.* (2001), Watt (1972).

Vernonia amygdalina Del.

Family: Asteraceae

Part/s Used	:	Roots, Stem and Leaves.
Utilization	:	Decoction useful in cough and pneumonia.
Active Constituents	:	Alkaloid, Inulin, saponin, starch and tannin isolated.
References	:	Gill *et al.* (1993), Karatela *et al.* (1991).

Veronica cinerea (L.) Less

**Syn.: *V. conyzoides* DC.; *V. rhomboidea* and *montana* Edgew.;
V. albicans DC.; *V. abbreviata* and *leptophylla* DC.; *V. physalifolia* DC.;
V. laxiflora Less.; *Conyza abbreviata, bellidifolia, cinerascens, incana,
linifolia, blegantula, ovata* Wall.; *Serratula cinerea* Willd.;
C. cinerea Linn.; *C.mollis* Willd.; *C. proliera* and *heterophylla* Lamk.**

Family: Asteraceae

Eng. and Sans. Names	:	*Eng.*: Ash coloured fleabane;
		Sans.: Sadodi, Sadori, Sahadevi.
Distribution	:	Throughout India ascending to 2,650m in Himalaya. Khasia Hills. Hills of Peninsula.
Part/s Used	:	Roots, Whole Plant and Seeds.
Utilization	:	Juice of roots taken for cough. Plant cures asthma and bronchitis. Seeds also useful in cough.
Active Constituents	:	β- amyrin acetate, β- amyrin benzoate, lupeol and its acetate, β- sitosterol, stigasterol and α- spinasterol isolated and–3β–acetoxy-19 -ene isolated from roots.
References	:	Ambasta (1986), Das and Aggarwal (1991), Kirtikar and Basu (1984), Rastogi and Mehrotra (1990, 1995), Watt (1972).

Viola biflora Linn.

Syn.: *V. manaslensis* Maekawa.

Family: Violaceae

Distribution	:	Temperate Himalaya.
Part/s Used	:	Whole Plant.

Utilization	:	Decoction of whole plant used as a beverage for curing pneumonia, cough and cold.
References	:	Gaur *et al.* (1983), Kirtikar and Basu (1984), Kumar (2002), Negi and Pant (1990), Sharma (2003).

Viola canescens Wall. *ex* Roxb.

Syn.: *V. serpens* Wall. var. *canescens* (Wall.) Hook. and Thoms.

Family: Violaceae

Part/s Used	:	Whole Plant.
Utilization	:	Leaves boiled for 30 min (preceded by a period of 3 days of drying) and used for treating cough and cold. Decoction of flowers also given in cough and cold.
References	:	Bennet (1987), Jain (1984), Lal *et al.* (1996), Negi *et al.* (1999), Rana *et al.* (1996), Rawat and Pangtey (1987).

Viola odorata linn.

Family: Violaceae

Eng., Sans. and H. Names	:	*Eng.*: Appel leaf, Bairnwort, Banwood, Blue violet;
		Sans.: Nilapushpa;
		H.: Banafshah, Vanapsa.
Distribution	:	Kashmir Himalaya: 1,700-2000m.
Part/s Used	:	Whole Plant.
Utilization	:	Rhizome as a component of herbal tea for relieving cough, flu and bronchitis.
Active Constituents	:	Rhizome contains glycoside–methyl salicylate, an alkaloid violine, a glycoside-violequarcitin and a saponin.
References	:	Kaul (1997), Kirtikar and Basu (1984), Watt (1972).

Viola pedata Linn.

Family: Violaceae

Eng. Name	:	*Eng.*: Bird's foot violet.
Part/s Used	:	Whole Plant.
Utilization	:	Decoction of plant used to loosen phleg in chest and for pulmonary problems.
Reference	:	Bhattacharjee (1998).

Viola pilosa Blume

**Syn.: *V.serpens* Wall. *ex* Ging., non Ridley; *V. serpens*
var. *glabra* Hook. and Thoms.; *V. palmaris* Buch.-Ham. *ex* Ging.**

Family: Violaceae

H. Name	:	*H.*: Banafsha.
Distribution	:	Throughout hilly regions.
Part/s Used	:	Whole Plant.
Utilization	:	Decoction of plant along with *Glycyrrhiza glabra* and *Adiantum lunulatum* given in cough as considered good bronchodilator and expectorant. Dried flowers boiled with tea used for cough and cold. Decoction prepared in combination with black pepper (5–10 ml, two times a day, for 3 days) beneficial in cough and cold.
Active Constituents	:	Rutin, violin and salicyclic acid reported from herb.
References	:	Ambasta (1986), Aswal (1996), Shah and Joshi (1971), Sharma (2003), Singh and Kumar (2000), Thakur *et al.* (1989).

Viola tricolor Linn.

Family: Violaceae

Eng. Names	:	*Eng.*: Flame flower, Garden gate, Pansy violet;
Distribution	:	Cultivated in India.
Part/s Used	:	Leaves and Flowers.
Utilization	:	Decoction of leaves and flowers given for asthma.
Active Constituents	:	Triterpene saponins (5.2 per cent) comprising of ursolic acid as aglycone and galactose or galacturonic acid, trans–caffeic, protocatechuic, gentisic, p–hydroxy–benzoic, 4- hydroxyphenylacetic, trans- and cis coumaric, vanillic and salicylic acids isolated along with 2 unidentified acids.
References	:	Ambasta (1986), Caius (2003), Kirtikar and Basu (1984), Rastogi and Mehrotra (1993).

Vitex negundo Linn.

**Syn.: *V. bicolor* Willd.; *V. arborea* Desf; *V. paniculata* Lam.
var. *incisa* Lamk.**

Family: Verbenaceae

Eng., Sans. and H. Names	:	*Eng.*: Indian privet;

		Sans.:Indrani, Nirgundi, Nitapushpa, Suvaha, Shephali;
		H.: Nisinda, Sambhalu, Shephali, Shivari.
Distribution	:	Throughout greater part of India ascending to 1,500 m in Himalaya.
Part/s Used	:	Leaves.
Utilization	:	Aqueous extract of leaves (5-10ml, 2 times a day, for a week) given orally for treatment of cough. Smoke of burnt leaves inhaled for curing asthma. Also, inhalation of vapours of leaves by boiling it in water considered good for severe cough and asthma.
Active Constituents	:	Fresh leaves yield essential oil (0.05 per cent) comprising limonene, camphene, b- phellandrene, caryophyllene, camphor, terpineol, cinnamaldehyde, β- sitosterol and vitexin.
References	:	Ambasta (1986), Jain (1996), Rajendran *et al.* (2001), Rastogi and Mehrotra (1995), Sharma (2003), Sinha and Sinha (2001), Watt (1972).

Vitis trifolia Linn.

Syn.: *Vitis carnosa* Wall; *Cayratia carmosa* Gagnep.; *V. incisa* Wall.

Family: Vitaceae

Eng., Sans. and H. Names	:	*Eng.*: Fleshy wild vine, Fox grape;
		Sans.: Banastha, Kandura, Saurasa, Tikshna;
		H.: Amalbel, Sufed–Sanbhalu.
Distribution	:	Throughout India.
Part/s Used	:	Leaves.
Utilization	:	Extract of leaves useful in tuberculosis.
Biological Activity	:	Extract exhibits inhibitory action against *Mycobacterium tuberculosis.*
References	:	Ambasta (1986), Kirtikar and Basu (1984), Nagendra Prasad and Abraham (1984), Watt (1972).

Wagatea spicata Dalz.

Syn.: *Caesalpinia mimosides* Heyne; *C.ferox* Hohen

Family: Fabaceae

Distribution	:	W. Ghats and Hills of W. Peninsula.

Part/s Used	:	Roots.
Utilization	:	Decoction of roots beneficial in pneumonia.
References	:	Ambasta (1986), Watt (1972).

Withania coagulans Dunal.

Syn.: *Puneeria coagulens* Stocks.; *Physalis somnifera* L.; *Withania somnifera* (L.) Dunal.

Family: Solanaceae

Eng., Sans. and H. Names	:	*Eng.*: Indian cheese maker, Indian Rennet;
		Sans.: Balaja, Vagini;
		H.: Akri, Punir.
Distribution	:	Tropical and Sub-tropical areas of India.
Part/s Used	:	Roots and Root-bark.
Utilization	:	Decoction of root-bark taken in asthma by Xosas. Root powder given orally to male patients of asthma and bronchitis (not to ladies as it acts as abortifacient).
Active Constituents	:	Root contains steroidal lactones, withaferins, withasomniferols, sitoindosides, ashwagandhine, withanine, curcohygrine, isopelletierine and visnamine.
Biological Activity	:	Respiratory stimulant activity found positive.
References	:	Ambasta (1986), Duke *et al.* (2002), Kirtikar and Basu (1984), Rastogi and Mehrotra (1991, 1993, 1995), Shah *et al.* (1983), Sharma (2003), Singh and Pandey (1998), Watt (1972).

Woodfordia fruticosa Kurz.

Syn.: *W. floribunda* Salisb.; *Lythrum fruticosum* Linn.

Family: Lythraceae

Eng., Sans. and H. Names	:	*Eng.*: Fire flame bush;
		Sans.: Agnijwala, Dhatri, Dhatupushpi;
		H.: Davi, Dhaura, Santha, Thawi.
Distribution	:	Wastelands in greater parts of India upto 1600m.
Part/s Used	:	Roots.
Utilization	:	Powdered roots given (5g, twice a day) for curing cough.
References	:	Ambasta (1986), Aminuddin and Girach (1996), Sharma (2003).

Wrightia tinctoria R.Br.

Syn.: *Nerium tinctorium* Roxb.

Family: Apocynaceae

Sans. and H. Names	:	*Sans.*: Hayamaraka, Svetakutaja;
		H.: Dhudhi, Khirmi, Inderjou.
Distribution	:	C. India. Peninsula.
Part/s Used	:	Leaves and Bark.
Utilization	:	Decoction of leaves and bark useful in asthma and bronchitis, respectively.
Active Constituents	:	β-sitosterol, β-amyrin and lupeol benzoate isolated from bark
References	:	Joshi *et al.* (1980), Kirtikar and Basu (1984), Rastogi and Mehrotra (1995), Shah *et al.* (1983), Watt (1972).

Zanthoxylum armatum DC.

Syn.: *Z.alatum* Roxb.; *Z.alatum* var. *planispinum* (Sieb. and Zucc.) Rehd. and E.H. Wils; *Z. planispinum* Sieb and Zucc.

Family: Rutaceae

Sans. and H. Names	:	*Sans.*: Andhka, Gandhalu,Tejohva;
		H.: Tejbal, Tejphal.
Distribution	:	Sub-tropical (ascending upto 1800m). Khasia Hills: 600-900m.
Part/s Used	:	Fruits.
Active Constituents	:	Monoterpenoid-3, 7–dimethyl–1–octane -3, 6, 7–triol, trans-cinnamic acid, nevadensin, umbelliferone, β-sitosterol and its glycoside isolated from fruits.
Remarks	:	Anti-asthmatic activity positive.
References	:	Anonymous (1999), Kirtikar and Basu (1984), Kumar and Dwivedi (2002), Rastogi and Mehrotra (1993, 1995).

Zanthoxylum rhetsa (Roxb.) DC.

Syn.: *Z. limonella* (Dennst.) Alston; *Z. budrunga* Wall. *ex* DC.; *Tipalia limonella* Dennst.; *Fagara budrunga* (Roxb.) DC.

Family: Rutaceae

Sans. and H. Names	:	*Sans.*: Ashvaghra, Sutejasi;
		H.: Badrang, Jaladhari, Pepuli.

Distribution	:	W. Peninsula from Coromandel and Konkan southwards.
Part/s Used	:	Fruits.
Utilization	:	Paste of powdered dry fruits of *Piper longum, P. nigrum* and *Z. rhetsa* (1:1:1) with a little water made into pills of about 3g each and given (2 pills, twice daily till cure) for asthma and bronchitis.
Active Constituents	:	Fruits afford alkaloids (0.24 per cent) and essential oil-mulliam oil.
References	:	Ambasta (1986), Chopra *et al.* (1956), Gogoi and Borthakur (2001), Kirtikar and Basu (1984), Watt (1972).

Zea mays Linn.

Family: Poaceae

Eng., Sans. and H. Names	:	*Eng.*: Indian corn, Maize corn;
		Sans.: Mahakaya, Makaya, Shikhalu, Yavanala;
		H.: Bhutta, Bhutta, Makai, Makka.
Distribution	:	Cultivated in India.
Part/s Used	:	Leaves and Inflorescence.
Utilization	:	Leaves useful in cough. Ash of the inflorescence used to cure whooping cough. Male flowers smoked to cure asthma and the ash left after smoking taken orally with water.
Active Constituents	:	Cellulose, tannin, lignin, inulin, di-O-(indole-3-acetyl)-myo-inositol and tri-O-(indloe-3-acetyl)-myoinositol isolated from kernels.
References	:	Ambasta (1986), Gill *et al.* (1993), Kirtikar and Basu (1984), Prakash and Mehrotra (1988), Rastogi and Mehrotra (1991), Singh and Pandey (1998).

Zingiber officinale Rosc.

Syn.: *Amomum zingiber* Linn.

Family: Zingiberaceae

Eng., Sans. and H. Names	:	*Eng.*: Ginger;
		Sans.: Anupama, Ardraka, Ardrashaka, Kandara, Machhaka, Mahija, Mulaja;
		H.: Ada, Adraka.
Distribution	:	Cultivated in hot and humid areas of S. India, Assam and H.P.

Part/s Used	:	Rhizome and Roots.
Utilization	:	Warm decoction of dried ginger relieves cough and asthma. 4-5 drops of rhizome juice also taken with honey twice, a day, for same. In case of dry and persistent cough, juice taken with a little 'sendha namak' (rock salt). For tuberculosis, pills made of powdered rhizome along with turmeric powder, jaggery and lime are prescribed (2 pills a day for 20 days). Among Mundas tribe of Chota Nagpur, pills made by frying fresh pounded roots with butter given for cure of cough (4 pills a day). Decoction prepared by boiling a few pieces of its rhizome and cinnamon leaf taken as a cough expectorant.
Active Constituents	:	Pungent principles are due to gingerols, shagols, dihydrogingerol,hexahydrocurcumin, gingerdiols, zingerone (rhizome), zerumone, zerumbodienone, humulene epoxide I and humulene epoxide 2 (root).
Remarks	:	Plant used in preparation of "Hub-gul-pista" for clearing respiratory system In Unani system.
References	:	Ambasta (1986), Kirtikar and Basu (1984), Jain and Tarafdar (1970), Kumar (2002), Panthi and Choudhray (2003), Rastogi and Mehrotra (1993), Sharma (2003), Singh (1996), Sinha and Sinha (2001), Thakur *et al.* (1989), Watt (1972), Yusuf *et al.* (2002).

Zingiber zerumbet (L.) Smith

Syn.: *Z. spurium* Kon.; *Amomum spurium* Gel.; *A. zerumbet* Willd

Family: Zingiberaceae

Eng., Sans. and H. Names	:	*Eng.*: Wild ginger;
		Sans.: Karpuraharidra;
		H.: Mahaaribach, Narkachur.
Distribution	:	Throughout India.
Part/s Used	:	Whole Plant and Rhizome.
Utilization	:	Rhizome as well as the whole plant employed as a hot remedy for cough and asthma.
Active Constituents	:	3″, 4″-O-diacetylafzelin along with zerumone epoxide and diferuloyl methane isolated from rhizome.
References	:	Ambasta (1986), Islam (1996), Jain and Tarafdar (1970), Kirtikar and Basu (1984), Tripathi and Goel (2001), Rastogi and Mehrotra (1993), Watt (1972).

Zizyphus mauritiana Ham.

**Syn.: *Zizyphus jujuba* Lour. non Mill.; *Z. sororia* Schult.;
Rhamnus jujuba Linn.**

Family: Rhamnaceae

Eng., Sans. and H. Names	:	*Eng.*: Chinese date, Indian cherry, Indian plum;
		Sans.: Ajapriya, Balashta, Kantaki, Rajakoli;
		H.: Baer, Ber, Beri.
Distribution	:	Throughout the greater part of India, in outer Himalaya upto 1,500m.
Part/s Used	:	Leaves, Seeds and Root-bark.
Utilization	:	Bark good for asthma. Pills made by pounding its roots (10g) in combination with those of 25g *Curculigo orchioides* and 5g *Zingiber officinale* (2 pills, twice a day) allay asthma.
Active Constituents	:	These are n-octacosanol, aliphitolic acid, eelin lactone, glucose, arabinose, 6-deoxy-L-talose, rutin, yuziphine, yuzirin, coclaurine, isoboldine, norisoboldine and asimilobine (leaves), spinosin (seeds).
References	:	Jain and Singh (1997), Kirtikar and Basu (1984), Rastogi and Mehrotra (1990, 1991, 1993, 1995), Tripathi and Kumar (2003), Watt (1972).

Zizyphus nummularia (Burm.) Wight and Arn.

**Syn.: *Z. microphylla* Roxb.; *Z. lotus* Lamk.; *Z. rotundifolia* Lam.;
Rhamnus nummularia Burm.f.**

Family: Rhamnaceae

Eng., Sans. and H. Names	:	*Eng.*: Wild jujube;
		Sans.: Ajapriya, Bhubadari, Bhukantaka;
		H.: Jhadiaber, Jhahrberi.
Distribution	:	Arid regions of Punjab, Gujarat, W. Rajputana and Cutch.
Part/s Used	:	Leaves.
Utilization	:	Leaves smoked for cough and cold. Also, its decoction with tea leaves useful in cough and cold.
Active Constituents	:	A dammarane saponin- Zizynummin isolated from leaves.
References	:	Ambasta (1986), Kirtikar and Basu (1984), Rastogi and Mehrotra (1993), Sinha and Pandey (1980), Watt (1972).

Epilogue

On the basis of available data, the following conclusions can be drawn:

Altogether 530 spp of plants belonging to 398 genera under 138 families have been recorded to be used by the tribals and indigenous communities of India against various lungs ailments, such as cough, cold, flu, asthma, bronchitis, pneumonia, tuberculosis, etc. (Appendix I). Of these the plant species belonging to angiosperms with 502 spp (439 spp of dicotyledons and 60 spp of monocotyledons) are the commonest to be used, followed by pteridophytes [16 spp, 12 genera: *Actinopteris* (1), *Adiantum* (5), *Chelianthes* (1), *Dicranopteris* (1), *Diplazium* (1), *Gleichenia* (1), *Lycopodium* (1), *Marsilea* (1), *Nephrolepis* (1), *Ophioglossum* (1), *Pellalaeo* (1), *Tectaria* (1)], gymnosperms (12 spp, 8 genera: *Abies* (1), *Cupressus* (1), *Ephedra* (1), *Gingko* (1), *Juniperus* (3), *Pinus* (3), *Taxodium* (1), *Taxus* (1)], fungi (2 spp, 2 genera: *Agaricus* (1), *Morchella* (1)], lichens (1 sp, I genus: *Usnea* (1)]. Of the angiospermous taxa, 439 dicotyledonous spp predominate (Figure 1; Appendix-I).

Analysis of the data presented in Appendix II clearly indicates that the most frequently used families are: Fabaceae (57 spp), Lamiaceae (31 spp), Asteraceae (29 spp), Zingiberaceae (18 spp), Solanaceae (16spp), Malvaceae (14 spp), Euphorbiaceae (13 spp), Apiaceae (11 spp), Rubiaceae (11 spp), Asclepiadaceae (10 spp), Brassicaceae (9 spp), Liliaceae (9 spp), Acanthaceae (8), Anacardiaceae (7), Arecaceae (7 spp), Lauraceae (7 spp), Myrtaceae (7 spp), Ranunculaceae (7 spp), Rosaceae (7 spp), Verbenaceae (7 spp), Araceae (6 spp), Cucurbitaceae (6 spp), Polypodiaceae (6 spp), Rutaceae (6 spp), Scrophulariaceae (6 spp), Violaceae (6 spp), Apocynaceae (5 spp), Menispermaceae (5 spp), Poaceae (5 spp), Amaranthaceae (4 spp), Boraginaceae (4 spp), Burseraceae (4 spp), Campanulaceae (4 spp), Capparidaceae (4 spp), Combretaceae (4 spp), Cupressaceae (4 spp), Pinaceae (4 spp), Piperaceae (4 spp),

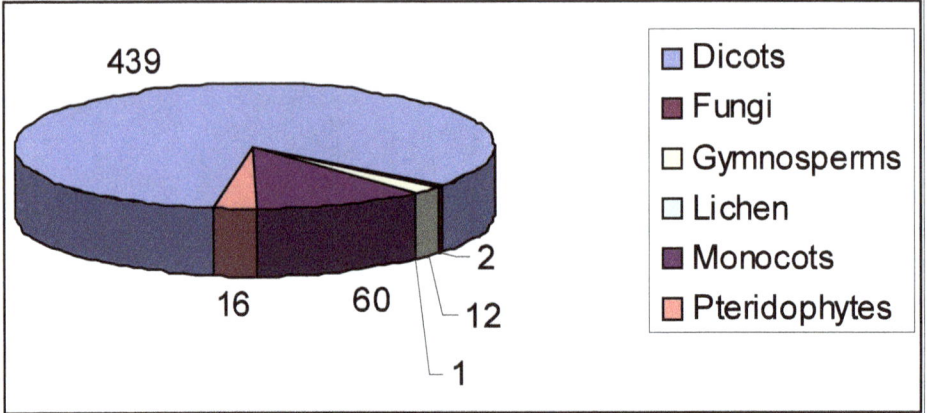

Figure 1: Total Number of Plants Under Various Divisions of Plant Kingdom for Healing Cold, Flu and Lungs Ailments

Polygalaceae (4 spp), Polygonaceae (4 spp), Sterculaceae (4 spp), Crassulaceae (3 spp), Elaeagnaceae (3 spp), Fagaceae (3 spp), Meliaceae (3 spp), Moraceae (3 spp), Orchidaceae (3 spp), Oxalidaceae (3 spp), Simaroubaceae (3 spp), Ulmaceae (3 spp), Zygophyllaceae (3 spp), etc. The remaining families are nonsignificant in uses. Obviously the size of the families has a relationship with these figures.

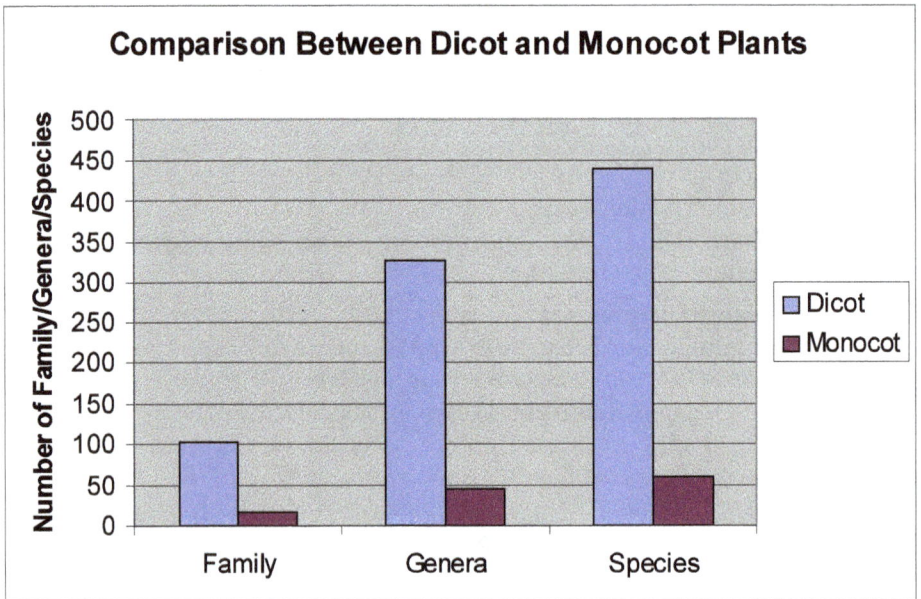

Figure 2: Comparison Between Dicot and Monocot Plants in Terms of Families, Genera and Species

Table 1: Index to Predominant Dicot Families of Plants Used for Healing Cold, Flu and Lungs Ailments with Relative% age in Taxa

Family	Total No. of Species	%age in Taxa	Family	Total No. of Species	% age in Taxa
Fabaceae	57	10.75%	Fagaceae	3	0.57%
Lamiaceae	31	5.85%	Meliaceae	3	0.57%
Asteraceae	29	5.47%	Moraceae	3	0.57%
Solanaceae	16	3.02%	Oxalidaceae	3	0.57%
Malvaceae	14	2.64%	Simaroubaceae	3	0.57%
Euphorbiaceae	13	2.45%	Ulmaceae	3	0.57%
Apiaceae	11	2.08%	Zygophyllaceae	3	0.57%
Rubiaceae	11	2.08%	Annonaceae	2	0.38%
Asclepiadaceae	10	1.89%	Aristolochiaceae	2	0.38%
Brassicaceae	9	1.70%	Bignoniaceae	2	0.38%
Acanthaceae	8	1.51%	Bombacaceae	2	0.38%
Anacardiaceae	7	1.32%	Cactaceae	2	0.38%
Lauraceae	7	1.32%	Caprifoliaceae	2	0.38%
Myrtaceae	7	1.32%	Caryophyllaceae	2	0.38%
Ranunculaceae	7	1.32%	Celastraceae	2	0.38%
Rosaceae	7	1.32%	Convolvulaceae	2	0.38%
Verbenaceae	7	1.32%	Dipterocarpaceae	2	0.38%
Cucurbitaceae	6	1.13%	Ebenaceae	2	0.38%
Rutaceae	6	1.13%	Gentianaceae	2	0.38%
Scrophulariaceae	6	1.13%	Leeaceae	2	0.38%
Violaceae	6	1.13%	Loganiaceae	2	0.38%
Apocynaceae	5	0.94%	Lythraceae	2	0.38%
Menispermaceae	5	0.94%	Myricaceae	2	0.38%
Amaranthaceae	4	0.76%	Ochnaceae	2	0.38%
Boraginaceae	4	0.76%	Oleaceae	2	0.38%
Burseraceae	4	0.76%	Pedaliaceae	2	0.38%
Campanulaceae	4	0.76%	Plantaginaceae	2	0.38%
Capparidaceae	4	0.76%	Podophyllaceae	2	0.38%
Combretaceae	4	0.76%	Portulacaceae	2	0.38%
Piperaceae	4	0.76%	Rhamnaceae	2	0.38%
Polygalaceae	4	0.76%	Salicaceae	2	0.38%
Polygonaceae	4	0.76%	Salvadoraceae	2	0.38%
Sterculiaceae	4	0.76%	Sapindaceae	2	0.38%
Crassulaceae	3	0.57%	Vitaceae	2	0.38%
Elaeagnaceae	3	0.57%			

**Table 2: Index to Predominant Monocot Families of Plants Used for
Healing Cold, Flu and Lungs Ailments with Relative% age in Taxa**

Family	Total No. of Species	% age in Taxa
Zingiberaceae	18	3.40%
Liliaceae	9	1.70%
Arecaceae	7	1.32%
Araceae	6	1.13%
Poaceae	5	0.94%
Orchidaceae	3	0.57%
Cyperaceae	2	0.38%

It is evident from Table 3 that species of genera like *Solanum* (8 spp), *Cassia* (7 spp), *Viola* (6 spp), *Adiantum* (5 spp), *Leucas* (5 spp), *Acacia* (4 spp), *Cinnamomum* (4 spp), *Curcuma* (4 spp), *Euphorbia* (4 spp), *Malva* (4 spp), *Piper* (4 spp), *Polygala* (4 spp), *Prunus* (4 spp), *Sida* (4 spp), *Ailanthus* (3 spp), *Alpinia* (3 spp), *Alysicarpus* (3), *Artemisia* (3 spp), *Datura* (3 spp), *Desmodium* (3 spp), *Ficus* (3 spp), *Indigofera* (3 spp), *Juniperus* (3 spp), *Lobelia* (3 spp), *Ocimum* (3 spp), *Pinus* (3 spp), *Saussurea* (3 spp), *Syzygium* (3 spp) and *Terminalia* (3 spp) predominate in uses.

Of the plants recorded in the present study, *Achyranthes aspera, Acorus calamus, Adhatoda vasica, Barleria cristata, Bergenia ciliata, Calotropis procera, Curcuma zedoria, Emblica officinalis, Myrica esculenta, Ocimum sanctum, Tinospora cordifolia* and *Viola odorata* are used for curing flu.

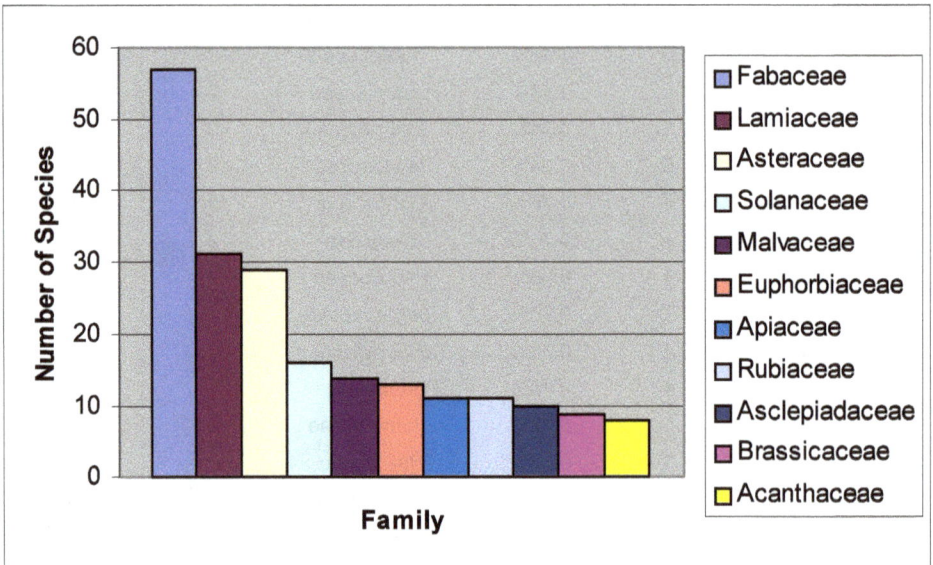

**Figure 3: Predominant Dicot Families of Plants Used for Healing
Cold, Flu and Lungs Ailments**

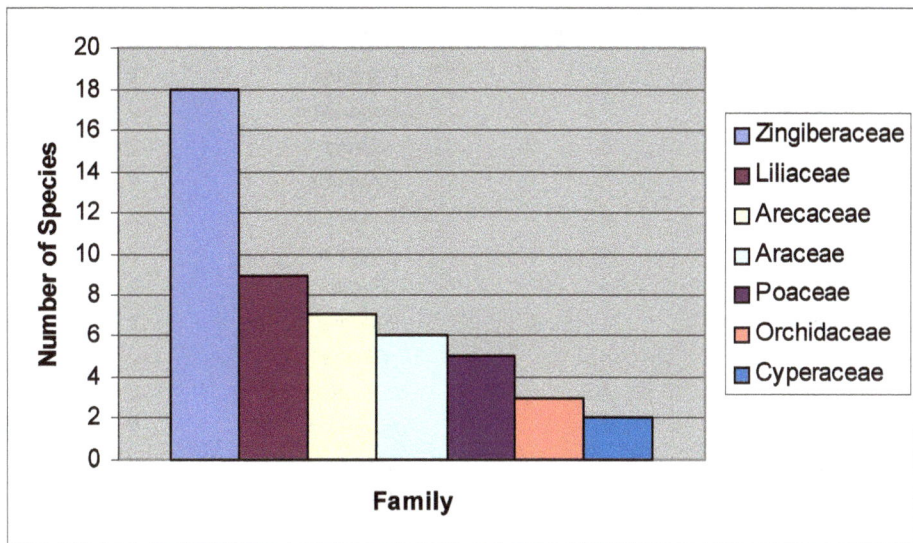

Figure 4: Predominant Monocot Families of Plants Used for Healing Cold, Flu and Lungs Ailments

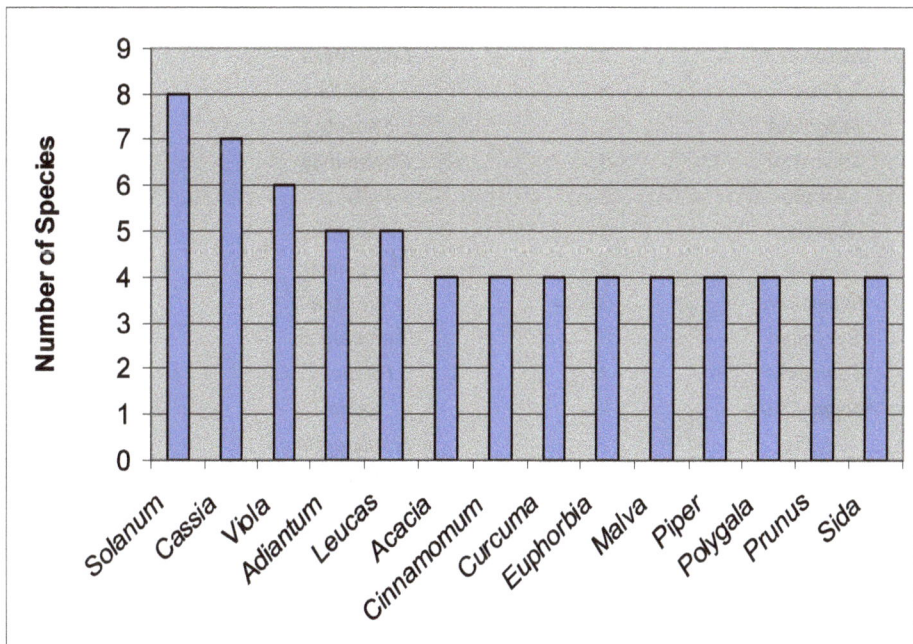

Figure 5: Predominant Genera of Plants Used for Healing Cold, Flu and Lungs Ailments

Table 3: Index to Predominant Genera for Healing Cold, Flu and Lungs Ailments

Name of Genus	No. of Species	Name of Genus	No. of Species
Solanum	8	Capparis	2
Cassia	7	Clerodendrum	2
Viola	6	Cordia	2
Adiantum	5	Diospyros	2
Leucas	5	Elaeagnus	2
Acacia	4	Fagonia	2
Cinnamomum	4	Ferula	2
Curcuma	4	Fritillaria	2
Euphorbia	4	Globba	2
Malva	4	Hedychium	2
Piper	4	Jasminum	2
Polygala	4	Kalanchoe	2
Prunus	4	Lactuca	2
Sida	4	Leea	2
Ailanthus	3	Luffa	2
Alpinia	3	Mentha	2
Alysicarpus	3	Mussaenda	2
Artemisia	3	Myrica	2
Datura	3	Ochna	2
Desmodium	3	Oldenlandia	2
Ficus	3	Opuntia	2
Indigofera	3	Phoenix	2
Juniperus	3	Phyllanthus	2
Lobelia	3	Pistacia	2
Ocimum	3	Plantago	2
Pinus	3	Polygonum	2
Saussurea	3	Portulaca	2
Syzygium	3	Pterocarpus	2
Terminalia	3	Rorippa	2
Abelmoschus	2	Salvia	2
Albizia	2	Stereospermum	2
Allium	2	Strychnos	2
Angelica	2	Tephrosis	2
Aristolochia	2	Tinospora	2
Atropa	2	Trema	2
Bambusa	2	Vanda	2
Brassica	2	Zanthoxylum	2
Caesalpinia	2	Zingiber	2
Calotropis	2	Zizyphus	2

It is needless to mention that the detailed clinical and experimental studies of these potential plant species are very much desired to evaluate the medical efficy of such plants. Moreover, such attempts will help in integration of traditional with modern system of medicine.

Literature Cited

Aditya, N.R. and Ghosh, R.B.1988. Useful angiosperms of Cooch Behar District, West Bengal. *J.Econ. Taxon. Bot.* **12**: 273-284.

Agrawal, V.S. 1986. *Economic Plants of India*. Bishen Singh, Mahendra Pal Singh, Dehradun and Kailash Prakashan, Calcutta.

Ambasta, S.P. (ed.) 1986.*The Useful Plants of India*.CSIR, New Delhi.

Aminuddin and Girach, R.D. 1991. Ethnobotanical studies on Bondo tribe of Distt. Koraput (Orissa), India. *Ethnobotany* **3**: 15-19.

Aminuddin and Girach, R.D. 1996. Native Phytotherapy among the Panda Bhuinya of Bonai Hills.*Ethnobotany* **8**: 66-70.

Anonymous. 1948-1976. *The Wealth of India, Raw Materials.* Vol. I-XI. Publications and Information Directorate, C.S.I.R., New Delhi.

Anonymous. 1994. *Ethnobiology in India- A Status Report.* Ministry of Environment and Forest, Govt. of India.

Anonymous. 1999. *The Ayurvedic Pharmacopoeia of India*. Part I. Vol. II. Government of India- Ministry of Health and Family Welfare. Department of ISM & H.

Ansari, A.A. 1991. Ethnobotanical notes on some plants of Khirsu, Pauri Garhwal, U.P. *Ethnobotany* **3**:105-106.

Arora, R.K. 1995. Ethnobotanical studies on plant genetic resources- national efforts and concern. *Ethnobotany* **1**: 125-136.

Arshad, Mohammed, Akbar, Ghulam and Iftikar, Qaiser. 1997. Medicinal plants of Cholistan desert. *J. Econ.Taxon. Bot.* **21**: 677-688.

Arya, K.R., Pande, P.C. and Prakash, Ved 1999. Ethnobotanical study on tribal areas of Almora District.- II. *Ethnobotany* **11**: 100-104.

Arya, K.R. and Prakash, Ved.1999.Ethnomedicinal study of a remote tribal area of Almora District. Survey Report-Part I. *J. Econ. Taxon. Bot*. **21**: 247-252.

Asolkar, L.V., Kakkar, K.K. and Chakre, O.J. 1992. *Second Supplement to Glossary of Indian Medicinal Plants with Active Principles*. Part-I (A-K), CSIR, New Delhi.

Aswal, B.S. 1996. Conservation of ethnomedicinal plant diversity of Garhwal Himalaya in India: 133-135. In Jain, S.K. (**ed.**) *Ethnobiology in Human Welfare*. Deep Publication, New Delhi.

Baily, C.J. and Day, C. 1989. Traditional plant medicines as treatment of diabetes. *Diabetes Care* **12**: 553-564.

Balasubramanium, P. and Prasad, S. Narendra. 1996. Medicinal plants among the Ivulars of Attappady and Boluvampatti forests in the Nilgiri biosphere reserve: 253-259 In Maheshwari, J.K. (**ed.**): *Ethnobotany in South- Asia*. Scientific Publishers, Jodhpur, India.

Banerjee, Archana1996. Medicinal uses of some flowers by Santhals of West Bengal: 314-317. In Maheshwari, J.K. (**ed.**) *Ethnobotany in South – Asia*. Scientific Publishers, Jodhpur, India.

Baruah, Parukutty and Sharma, G.C. 1987. Studies on the medicinal uses of plants by the North-East tribe-III. *J. Econ. Tax. Bot*. **11**: 71-76.

Bennet, S.S.R. 1987. *Name Changes in Flowering Plants of India and Adjacent Regions*. Triseas Publishers, Dehradun, India.

Bhalla, N.P., Sahu, T.R., Mishra, G.P. and Dakwale, R.N.1982. Traditional plant medicines of Sagar Distt., M.P, India. *J. Econ. Tax. Bot*. **3**: 23-32.

Bhalla, Suman, Patel, J.R. and Bhalla, N.P. 1996. Ethnomedicinal observation on some Asteraceae of Bundelkhand region, M.P.: 175-178. In Maheshwari, J.K. (**ed.**) *Ethnobotany in South – Asia*. Scientific Publishers, Jodhpur, India.

Bhandary, M.J. and Chandrashekhar, K.K. 2002. Glimpses of ethnic herbal medicines of coastal Karnataka. *Ethnobotany* **14**:1-12.

Bhandary, M.J., Chandrashekhar, K.K and Kaveriappa, K.M. 1996. Ethnobotany of Gowlis of Uttara Kannada Distt., Karnataka: 244-249. In Maheshwari, J.K. (**ed.**) *Ethnobotany in South- Asia*. Scientific Publishers, Jodhpur, India.

Bhatnagar, P. 1997. Marketing and trade of medicinal plants. *Sustain for*. **11**: 4-11.

Bhatt, D.C., Mehta, D.R., Mehta, S.K. and Parmar, R.P. 2002. Studies on some ethnomedicinal plants from Talaja Taluka of Bhavnagar Distt., Gujarat. In Trivedi, P.C. (**ed.**) *Ethnobotany*. Aavishkar Publishers Distributors, Jaipur 302003 (Raj) India.

Bhatt, D.C. and Mitaliya, K.D. 1999. Ethnomedicinal plants of Victoria park of Bhavnagar, Gujarat, India. *Ethnobotany* **11**: 81-84.

Bhatt, D.C., Mitaliya, K.D. and Mehta, S.K. 2001.Some sacred plants of Gujarat and their medicinal uses. *Ethnobotany* **13**:146-149.

Bhatt, D.C., Mehta, S.K. and Mitaliya, K.D.1999. Ethnomedicinal plants of Shetrunjava hills of Palitana, Gujarat. *Ethnobotany* **11**: 22-25.

Bhatt, D.C., Mitaliya, K.D., Mehta, S.K. and Joshi, P.N. 2002. Note on some ethnomedicinal plants of Pachchham hills of Kachch District, Gujarat. *Ethnobotany* **14**: 34-35.

Bhatt, R.P. and Sabnis, S.D. 1987. Contribution to the ethnobotany of Khedbrahma region of North-Gujarat. *J. Econ. Tax. Bot.* **9**: 139-145.

Bhattacharjee, S.K.1998. *Handbook of Medicinal Plants*. Pointer Publishers, New Delhi.

Bhattacharjee, S.K. 2000. *Handbook of Aromatic Plants*. Pointer Publishers, New Delhi.

Bhattacharjee, S.K. 2001. *Handbook of Medicinal Plants*. Pointer Publishers, Jaipur.

Bhogaonkar, P. Y. & Devarkar, V. D. 2002. Some unique ethnomedicinal plants of Korkus of Melghat Tiger Reserve (Maharashtra). *Ethobotany* 14: 34-35.

Brahmam, M. and Saxena, H.O.1990. Ethnobotany of Gandhamardan Hills- Some noteworthy folk medicinal uses. *Ethnobotany* **2**: 71-74.

Caius, J.F. 2003. The medicinal and poisonous ferns of India: 34-54 In Caius, J.F. (**ed**.) *The Medicinal and Poisonous Plants of India*. Scientific Publishers, Jodhpur.

Chakraborhty, M. and Hazra, D. 2003. Aquatic monocots of West Bengal–Family Hydrocharitaceae, Xyridaceae and Alismataceae. *J. Econ. Taxon. Bot.* **27**: 1140-1150.

Chandler, R.F. and Hooper, S.N. 1982. Ethnobotany and phytochemistry of Yarrow. *Econ. Bot.*

Chandrasekar, K. & Srivastava, S. K. 2003. Ethnomedicinal studies in Pin valley national park, Lahaul-Spiti, Himachal Pradesh. *Ethnobotany* **15**: 44-47.

Chaudhary, U.S. and Hutke, Varsha. 2002. Ethno-medico-botanical information on some plants used by Melghat tribes of Amravati District, Maharashtra. *Ethnobotany* **14**: 100-102.

Chatterjee, A. and Pakrashi, S.C. (**ed**.) 1997. *The Treatise on Indian Medicinal Plants*. Vol. 1-5. CSIR, New Delhi.

Chaurasia, O.P, Singh, Brahma and Sareen, S.K.1999. Ethno-medico-botanical survey of Nubra Valley. *J. Econ. Taxon. Bot.* **23**: 167-173.

Chhetri, R.B.1994. Further observations on ethnomedico-botany of Khasi hills in Meghalaya, India. *Ethnobotany* **6**: 33-36.

Chopra, I.C., Khajuria, B.N. and Chopra, C.L. 1957. Antibacterial properties of volatile principles of *Alpinia galanga* and *Acorus calamus*. *Antibiotics Chemother* **7**: 378.

Chopra, R.N., Chopra, I.C. and Varma, B.S. 1968. *Supplement to Glossary of Indian Medicinal and Allied Plants*. CSIR, New Delhi.

Chopra R.N., Chopra I.C. and Varma B.S. 1980. *Supplement to Glossary of Indian Medicinal Plants*. Publications and Information Directorate, New Delhi.

Chopra, R.N., Nayar, S.L. and Chopra, I.C. 1956. *Glossary of Indian Medicinal Plants*. CSIR, New Delhi.

Dagar, H.S. and Dagar, J.C. 1996. Ethnobotanical studies of the Nicobarese of Chowra island of Nicoar group of islands.: 381-388. In Maheshwari, J.K. (**ed.**) *Ethnobotany in South – Asia*.Scientific Publishers, Jodhpur, India.

Das, Doli and Agarwal, V.S.1991. *Fruit Drug Plants of India*. Kalyan Publishers, New Delhi-Ludhiana.

Das, M.B. and Sharma, T.C. 1958-59. Traditional methods of treatment of leprosy among Mikris. *Adibasi* **3**: 25-30.

Das, P.K. and Misra, M.K.1988. Some medicinal plants among Kondhas around Chandrapur (Koraput). *J. Econ. Taxon. Bot*.**12**: 63-109.

Das, S.N. 1997. A study on the ethnobotany of Karauli and Sawai Madhopur Distt, Rajasthan. *J. Econ. Taxon. Bot*. **21**: 587-605.

Das, S.N., Janardhanan, K.P. and Roy, S.C. 1983. Some observations on the ethnobotany of the tribes of Totopara and adjoining area in Jalpaiguri Distt., W. Bengal. *J. Econ. Tax. Bot*. **4**: 453-474.

Desai, B.S. and Jasraj, Y.T. 2003. Herbal resources available for commonest disease-diabetes.: 35-43. In Khan, I. A. and Khanum, A. (**ed**.) Role of iotechnology in Medicinal and Aromatic Plants, Vol. VIII. Ukaaz Publication, Hyderabad.

Deshmukh, V.R., Muratkar, G.D. and Rothe, S.P.1999. Prilimnary observation on medicinal and economically important leguminous plant species from Amravati tehsil. *J. Econ. Taxon. Bot*. **23**: 283-289.

Dey, K.L. 1896. *The Indigenous Drugs of India*. Pharma Primlane the Chronica, New Delhi, India.

Dixit, R.D. and Pal, D.C. 1996. Role of household remedy in primary health care: 108-109. In Jain, S.K. (**ed**.) *Ethnobiology in Human Welfare*. Deep Publication, New Delhi.

Duke, James A., Bogenschutz-Godwi, M.J., Judi, duCellier and Duke, P.K. 2002. *Handbook of Medicinal Herbs*. CRC Press Boca Raton London. New York. Washington D.C.

Duke, J.A., Bogenschutz-Godwi, M.J., Judi duCellier & Duke, P.K. 2002. *Handbook of Medicinal Herbs*. CRC Press, New York.

Fransworth, N.R. 1985. Plants and modern medicine: Where science and folklore meet. *Eastern Pharmacist* **28**: 33-36.

Garcia da Orta. 1563. *Colloquious Des Simples Drogas- The Consassedicinai da India*, Goa.

Gaur, R.D., Semwal, J.K. and Tiwari, J.K. 1983. A survey of high altitude medicinal plants of Garhwal Himalaya. *Bull. Medico-Ethnobot. Res*. **4**: 102-116.

Girach, R.D., Aminuddin, Brahmam, M. and Mishra, M.K. 1997. Observations on ethnomedicinal plants of Bhadrad Distt., Orissa, India. *Ethnobotany* **9**: 44-46.

Gill, L.S., Idu, M. and Ogbor, D.N. 1997. Folk medicinal plants: practices and beliefs of the Bini people in Nigeria. *Ethnobotany* 9: 1-5.

Gill, L.S. and Nyawuame, H.G.K. 1994. Leguminosae in ethnomedicinal practices of Nigeria. *Ethnobotany* 6: 51-64.

Gill, L.S., Nyawuame, H.G.K., Esezobor, E.I. and Osagie, I.S. 1993. Nigerian folk medicine: practices and beliefs of Esan people. *Ethnobotany* 5:129-142.

Goel, A.K. and Bhattacharya, U.C. 1981. A note on some plants found effective in treatment of jaundice (Hepatitis). *J. Econ. Tax. Bot.* 2: 157-159.

Gogoi, R. and Borthakur, S.K. 2001. Notes on herbal recipes of Bodo tribe in Kamrup District, Assam. *Ethnobotany* 13:15-23.

Goud, P.Saiprasad. and Pullaiah, T.1996.Ethnobotany of Kurnool District, Andhra Pradesh: 410-412. In Jain, S.K. (**ed.**) *Ethnobiology in Human Welfare*. Deep Publications, New Delhi.

Greenish, Henry. G. 1999. *Materia Medica*. Scientific Publishers (India), Jodhpur.

Grover, J.K., Yadav, S. and Vats,V. 2002. Medicinal plants of India with anti-diabetics potential. *J. Ethnopharmacol* 81: 81-100.

Gupta, R. 1986. Prospects off cultivation off plants used in Ayurvedic medicine into agriculture and agro-forestry systems in India. *Prc. Silver Jublee Celebration Nehru Ayurvedic Medicinal Plants Garden and Herbarium. Pune*: 321-333.

Gupta, S.P. 1997. Native medicinal uses of plants by Asurs of Natarhat plateau (Bihar).: 103-116. In Jain, S.K. (**ed**) *Contribution to Indian Ethnobotany*. 3rd Edition. Scientific Publishers. Jodhpur/India.

Harborne, Jeffrey B. and Baxter, Herbert 2001. *Chemical Dictionary Of Medicinal Plants*. John Wiley and Sons Ltd. Chichester, New York. Weinheim. Brisbane. Singapore. Toronto.

Hembrom, P.P.1991. Tribal medicine in Chotanagpur and Santhal Parganas of Bihar, India. *Ethnobotany* 3: 97-99.

Islam, M.1983. Utilisation of Certain ferns and fern allies in the North-Eastern region of India. *J. Econ. Tax. Bot.* 4: 861-867.

Islam, M.1996. Ethnobotany of certain underground Parts of North-Eastern region, India: 338-343. In Maheshwari, J.K. (**ed.**) *Ethnobotany in South – Asia*. Scientific Publishers, Jodhpur, India.

Jadeja, B.A. 1999. Plants used by the tribe Rebury in Barda Hills of Gujarat *Ethnobotany*11: 42-46.

Jain, S.K. 1964. Native plant remedies for snake-bite among adivasis of Central India. *Ind. Med. J.* 57: 307-369.

Jain, S.K. 1967. Ethnobotany- Its scope and study. *Indian Mus. Bull.* 2: 39-43.

Jain, S.K.1968. *Medicinal Plants*. National Book Trust, India.

Jain, S.K. 1986. Ethnobotany. *Interdiscipl. Sci. Rev.* II: 285-292.

Jain, S.K. 1995. Ethnobotanical diversity in Zingibers of India. *Ethnobotany* 1: 83-88.

Jain, S.K. 1997. *Contribution to Indian Ethnobotany*. Scientific Publishers. Jodhpur.

Jain, S.K. 2002. *Bibiliography of Indian Ethnobotany*. Scientific Publication, Jodhpur.

Jain, S.K. & Sikarwar, R.L.S. 1998. Some Indian plants used in Latin American ethnomedicine. *Ethnobotany* **10** (1 & 2): 61-65.

Jain, S.K. & Tarafdar, C.R. 1970. Medicinal plant lore of the Santals. A revival of P.O. Boddings's work. *Econ. Bot.* 24: 241-278.

Jain, S.K., Fernandes, V. Froes., Lata, Sneh and Ayub, A. 1995. Indo- Amazonicm etnobotanic connections–similar uses of some common plants. *Ethnobotany* **7**: 29-37.

Jain, S.K., Mudgal, V., Banerjee, D.K., Guha, A., Pal, D.C. and Das, D. 1984. *Bibiliography of Ethnobotany*. B.S.I., Howrah.

Jain, S.K. and Saklani, Arvind. 1992. Cross cultural ethnobotanical studies on Northeast India. *Ethnobotany* **4**: 25-38.

Jain, S.K., Sirkarwar, R.L.S. and Pathak, V.1997.Ethnobotanic aspects of some plants in Latin America. *Ethnobotany* **9**: 16-23.

Jain, S.K. and Srivastava, S. 2001. Indian ethnobotanical literature in last two decades- A graphic review and future directions. *Ethnobotany* **13**: 1-8.

Jain, S.K. and Tarafdar, C.R. 1970. Medicinal plantlore of the Santals. A revival of P.O. Bodding's work. *Econ. Bot.* **24**: 241-278.

Jain, S.P. 1984. Ethnobotany of Morni and Kalesar (Ambala–Haryana). *J. Econ. Tax. Bot.* **5**: 809-813.

Jain, S.P. 1996. Ethno-medico-botanical survey of Chaibasa, Singhbhum District, Bihar: 403-407 In Maheshwari, J.K. (**ed.**) *Ethnobotany in South – Asia*. Scientific Publishers, Jodhpur, Ind.

Jain, S.P. & Singh, S.C. 1997. An ethno-medico-botanical survey of Ambikapur district, M.P: 83-91. In Jain, S.K. (**ed**.*). Contribution to Indian Ethnobotany* 3ʳᵈ ed. Scientific Publishers, Jodhpur, India.

Jamir, N.S. 1997. Ethnobiology of Naga tribe in Nagaland: I-medicinal herbs. *Ethnobotany* **8**: 101-104.

Jha, R.R. and Varma, S.K.1996. Ethnobiology of Sauria Paharias of Santhal Pargana, Bihar: I. medicinal plants. *Ethnobotany* **8**: 31-35.

Joshi, M.C., Patel, M.B. and Mehta, P.J. 1980. Some folk medicine of Dangs, Gujarat state. *Bull. Medico. Ethnobot. Res.* **1**: 8-24.

Joshi, P. 1982. An ethnobotanical study of Bhils–A preliminary survey. *J. Econ. Tax. Bot.* **3**: 257-226.

Katewa, S.S. and Arora, Asha. 1997. Some plants in folk medicine of Udaipur District (Rajasthan). *Ethnobotany* **9**: 48-51.

Katewa, S.S., Chaudhary, B.L., Jain, A.P. & Thakha, H.K. 2001. Some plants in folk medicine of Rajasamand District (Rajasthan). *Ethnobotany* **13**: 129-134.

Kapur, S.K. 1996. Traditionally important medicinal plants of Bhaderwah Hills–Jammu Province III: 62-69. In Maheshwari, J.K. (**ed.**) *Ethnobotany in South – Asia*. Scientific Publishers, Jodhpur, India.

Kapur, S.K., Nanda, S. and Srivastava, T.N. 1996. Ethnobotanical uses of RRL-Herbarium III.: 50-55. In Maheshwari, J.K. (**ed.**) *Ethnobotany in South – Asia*. Scientific Publishers, Jodhpur, India.

Karatela,Y.Y., Omokafe,B.A. and Nurani, M.A. 1991. Survey of fold medicinal plants used by the Okpameri tribe in Akoko Edo area of Bendel state of Nigeria. *Ethnobotany* **3**: 51-55.

Karnick, C.R., Tiwari, K.C., Majumdar, R. and Bhattacharjee, S. 1981. Newer ethnobotanical and folklore studies of some medicinal plants of Gauhati and surrounding areas. *Nagarjun.* **24**: 240-245.

Kaul, M.K. 1997. *Medicinal Plants of Kashmir and Ladakh*. Indus Publishing Company, New Delhi.

Khan, S.S. and Chaghati, S.A. 1982. Ethnobotanical Studies on some plants used for curing skin affections. *Anc. Sci. Life* **1**: 236-238.

Khanna, K.K. & Kumar, R. 2000. Ethnomedicinal plants used by the Gujjar tribe of Saharanpur district, Uttar Pradesh. *Ethnnobotany*. **12** (1 & 2): 17-22.

Khanna, K.K. and Kumar, Ramesh. 2002. Ethnomedicinal plants used by the Gujjar tribe of Saharanpur Distt., U.P. *Ethnobotany* **12**:17-22.

Khanna, K.K., Shukla, G. and Mudgal, V. 1996. New traditional medicinal uses of plants from Jalaun Distt., U.P.: 108-111. In Maheshwari, J.K. (**ed.**) *Ethnobotany in South Asia*. Scientific Publishers, Jodhpur, India.

Khanna, K.K., Srivastava, P.K. and Mudgal, V. 1996. Noteworthy medicinal plant uses from rural folklore of Raebareli Distt., U.P: 198-122. In Maheshwari, J.K. (**ed.**) *Ethnobotany in South Asia*. Scientific Publishers, Jodhpur, India.

Khare, C.P. 2004. *Encyclopedia of Indian Medicinal Plants*. Rational western therapy, Ayurvedic and other traditional usage, Botany. Springer- Verlage Berlin Heidelberg New York.

Kharkongor, P. and Joseph, J. 1981. Folklore medico-botany of rural Khasi and Jiantia tribes in Meghalaya: 124-136. In Jain, S.K. (**ed**.) *Glimpses of Indian Ethnobotany*. Oxford and IBH Publishing Co., New Delhi.

Kirtikar, K.R. and Basu, B.D. 1984. *Indian Medicinal Plants*. Vols. I- 1V. Lalit Mohan Basu, Allahabad.

Kothari, M.J. and Londhe, A.N. 1999. Ethnobotany in human health care of Chikhaldara, Amaravati District in Maharashtra state, India. *J. Econ. Taxon. Bot.* **23**: 273-281.

Kothari, M.J. and Moorthy, S. 1996. Ethnobotany in human welfare of Raigad Distt. In Maharashtra State, India: 403-407. In Jain, S.K. (**ed.**) *Ethnobotany In Human Welfare*. Deep Publications, New Delhi.

Kothari, M.J. and Rao, K.M. 1999. Ethnobotanical studies of Thane District, Maharashtra. *J. Econ. Taxon. Bot.* **23**: 265-272.

Kshirsagar, R.D. and Singh, N.P. 2000. Less known ethnomedicinal uses of plants in Coorg Distt. of Karnataka state, southern India. *Ethnobotany* **12**: 12-16.

Kshirsagar, Sanjay.R., Parabiam, M.H. and Reddy, M.N. 2003. Ethnobotany of coastal areas in south Gujarat. *Ethnobotany* **15**: 60-63.

Kumar, Sudhir. 2002. *The Medicinal Plants of North-East India*. Scientific Publishers. Jodhpur. India.

Kumar, Sushil and Dwivedi, Samresh. 2000. Agrotechnological gaps and focussed interventions for commercial cultivation of various medicinal and aromatic Plants: 111-124. In Nautiyal A.R., Nautiyal M.C. and Purohit A.N. (**ed.**) *Harvesting Herbs. Medicinal And Aromatic Plants. An Action Plan For Uttarkhand*. Bishen Singh Mahendra Pal Singh, Dehradun.

Kumar, S. & Singh, A. K. 2000. Research and development in medicinal and aromatic plants: 77-97. In Nautiyal A.R., Nautiyal M.C. and Purohit A.N. (**ed.**): *Harvesting Herbs. Medicinal And Aromatic Plants. An Action Plan For Uttarkhand*. Bishen Singh Mahendra Pal Singh, Dehradun.

Kumar, T. Dharma Chandra and Pullaiah, T. 1999. Ethno-medicinal uses of some plants of Mahabubnagar District, Andhra Pradesh, India. *J. Econ. Tax. Bot.* **23**: 341-345.

Kumar, V. & Jain, S.K. 1998. A contribution to ethnobotany of Surguja district in Madhya Pradesh, India. *Ethnobotany* **10** (1 & 2): 89-96.

Kumar, Vijaya. R. and Pullaiah, T. 1998. Medicinal plants used by the tribals of Prakasam District, A.P. *Ethnobotany* **10**: 97-102.

Kumar, Vivek. and Rao, R.R. 2001. Some plant beverages used in traditional medicine. *Ethnobotany* **13**: 36-39.

Lal, Brij, Vats, S.K., Singh, R.D. and Gupta, Akshey K. 1996. Plants used as ethnomedicinal and supplementary food by Gaddis of Himachal Pardesh, India: 384-387. In Jain, S.K. (**ed.**) *Ethnobotany in Human Welfare*. Deep Publication, New Delhi.

Lalramnghinglova, J.H. 1996. Ethnobotany of Mizoram- A preliminary survey: 439-459. In Maheshwari, J.K. (**ed.**) *Ethnobotany in South Asia*. Scientific Publishers, Jodhpur, India.

Lalramnghinglova, J.H. 1999. New records of ethnomedicinal plants from Mizoram. *Ethnobotany* **11**: 57-64.

Lalramnghinglova, J.H. and Jha, L.K. 1997. Ethnomedicine from Mizoram- North-East India. *Ethnobotany* **9**:105-111.

Lal, S.D. and Yadav, B.K. 1983. Folk medicine of Kurukshetra District (Haryana), India.*Econ. Bot.* **37**: 299-305.

Maheshwari, J.K., Painuli, R.M. and Dwivedi, R.P.1997. Notes on ethnobotany of Oraon and Korwa tribes of Madhya Pradesh: 67-82. In Jain, S.K. (**ed**.) *Contribution to Indian Ethnobotany*. Third Edition. Scientific Publishers.

Maheshwari, J.K. and Singh, J.P. 1987. Traditional Phytotherapy amongst the Kol Tribe of Banda Disrict, U.P. *J. Econ. Tax. Bot.* **9**:165-171.

Malhotra, C.L. and Balodi, B. 1984. Wild medicinal plants in the use of Johri tribals. *J. Econ. Tax. Bot.* **5**: 841-843.

Malhotra, S.K. and Moorthy, S. 1973. Some useful and medicinal plants of Chandrapur Distt. (Mahrashtra). *Bull. Bot. Surv. India.* **5**: 223-226.

Maliya, S. D and Singh, K.K. 2003. Some new or less known folk medicines of Bahraich Distt., U.P. *Ethnobotany* **15**:132-135.

Manandhar, N.P. 1990. Folklore medicine of Chitwan District, Nepal. *Ethnobotany* **2**: 31-38.

Manandhar, N. P. 1996. Traditional practice for oral health care in Nepal. *J. Econ. Taxon. Bot.* Additional Series 12: 408-413.

Mandal, S. K. and Basu, S.K.1996. Ethnobotanical studies among some tribals of Nilgiri District, Tamil Nadu: 268-271. In Maheshwari, J.K. (**ed.**) *Ethnobotany in South Asia*. Scientific Publishers, Jodhpur, India.

Maya, S., Prameela, Kumari, S. and Menon, V. Sarojini. 2003. Etnobotanical notes on the flora of sacred tanks of Kerala. *Ethnobotany* **15**: 55-59.

Megoneitso and Rao, R.R. 1983. Ethnobotanical studies in Nagaland-4.62 Medicinal plants used by Angami Naga. *J. Econ. Tax. Bot.* **4**: 167-172.

Mehrotra, Shanta, Rawat, A. K. S., Singh, H.K. and Shome,Usha 1995. Standardization of popular ayurvedic adaptogenic preparation chyavanprash and ethnobotany of its ingredients. *Ethnobotany* **7**:1-15.

Misra, L.N. and Kumar, S. 2000. Indian medicinal plants as the potential source of therapeutic agents: 13-27. In Nautiyal A.R., Nautiyal M.C. and Purohit A.N. (**ed**.) *Harvesting Herbs*-2000. *Medicinal And Aromatic Plants. An Action Plan For Uttarkhand*. Bishen Singh Mahendra Pal Singh, Dehradun.

Mishra, R. and Dixit, R.D.1976. Studies on ethnobotany of some less known medicinal plants of Ajmer forest division. Rajasthan. *Nagarjun.* **19**: 20-22.

Mudgal, V. and Pal, D.C. 1980. Medicinal plants used by tribals of Mayurbhanj (Orissa). *Bull. Bot. Surv. India.* **22**: 59-62.

Mukherjee, A., Mukherjee, S. and Mukherjee, M. 1986. Less known medicinal plants of Rubiaceae in Darjeeling hills, W.Bengal (India). *J. Appl. and Pure Bot.* **1**: 86-91.

Mukherjee, Ambarish and Namhata, Debashis. 1990. Medicinal plantlore of the tribals of Sundargarh District, Orissa. *Ethnobotany* **2**: 57-60.

Nadkarni, K.M. 1926. *Indian Materia Medica*. Popular Book Depot, Bombay.

Nagendra Prasad, P. and Abraham, Z. 1984. Ethnobotany of Nayadis of N. Kerala. *J. Econ. Tax. Bot.* **5**: 41-48.

Nayak, P.K., Nayak, R.K. & Choudhury, B.P. 2003. A check list of medicinal plants of Kalahandi district in Orissa: 519-532. In Singh, V. & Jain, A.P. (**ed.**). Vol.-²². *Ethnobotany and Medicinal Plants of India and Nepal.* Scientific Publishers, Jodhpur, India.

Negi, K.S. and Pant, K.C. 1990. Notes on ethnobotany of the Gangwal – a tribe of Garhwal Himalaya. *Ethnobotany* **2:** 81-85.

Negi, K.S., Tiwari, J.K.and Gaur, R.D. 1985. Economic importance of some common trees in Garhwal Himalaya: An ethnobotanical study. *Indian. J. Forestry* **8**: 276-289.

Negi, K.S., Tiwari, J.K., Gaur, R.D.and Pant, K.C. 1993. Notes on ethnobotany of five Districts of Garhwal Himalaya, U.P, India. *Ethnobotany* **5**: 73-81.

Negi, K.S., Tiwari, J.K., Gaur, R.D. and Pant, K.C.1999. Ethnobotanical notes on the flora of Har-Ki-Doon (Distt. Uttarkashi) Garhwal Himalaya, Uttar Pradesh, India. *Ethnobotany* **11**: 9-17.

Nunez, Diego Rivera and Castro, Concepcion Oon De. 1995. Medicinal plants and a multipurpose complex mixture sold in the market of Funchal (Island of Madeira, Portugal). *Ethnobotany* **7**: 75-82.

Nwosu, M.O. 2001. Ethnobotanical studies on some pteridophytes of southern Nigeria. *Ethnobotany* **13:** 29-35.

Painuli, R.M. and Maheshwari, J.K.1996. Some interesting ethnomedicinal plants used by Sahariya tribe of Madhya Pradesh: 179-185. In Maheshwari, J.K. (**ed.**) *Ethnobotany in South- Asia.* Scientific Publishers, Jodhpur, India.

Panda, S. 1996. Plant use- A recent perspective from eastern Himalaya: 332-337. In Maheshwari, J.K. (**ed.**) *Ethnobotany in South Asia.* Scientific Publishers, Jodhpur, India.

Pandey, Bhawna and Pandey, P.C.1999. Ethnobotanical studies on gymnospermic plants of Kumaon Himalaya. *J. Econ. Taxon. Bot.* **23**: 253-256.

Pandey, Indu Bhushan 2003. Some traditional herbal household remedies used in and around Kanpur city (U.P), India. *Ethnobotany* **15**: 129-131.

Pandey, N. K., Joshi, G. C., Mudaiya, R. K., Tewari, V. P. & Kewari, K. C. 2003. Management and Conservation of Medicinal Orchids of Kumaon and Garhwal Himalaya: 114-118. In Singh, V. & Jain, A. P. (**eds.**): *Ethnobotany and Medicinal Plants of India and Nepal Vol. I.* Scientific Publisher, Jodhpur (India).

Panthi, Mohan P. and Chaudhary, Ram P. 2003. Ethnomedicinal plant resource of Arghakhanchi District, West Nepal. *Ethnobotany* **15**: 71-86.

Prakash, Ved and Mehrotra, B.N. 1987. Ethnobotanical studies on the flora of Khandala, Maharashtra State. *J. Econ. Tax. Bot.* **9**: 205-208.

Prakash, Ved. and Mehrotra, B.N.1988. Unrecorded traditional medicines–I. *Anc. Sci. Life.* **3**: 110 -112.

Prakash, Ved. and Mehrotra, B.N.1991. Ethno-medicinal uses of some plants among Garos of Meghalya. *Ethnobotany* **3**: 41-45.

Prasad, N.P. & Abraham Z. 1984. Ethnobotany of Nayadis of North Kerela. *J. Econ. Tax. Bot.* **5**: 41-48.

Prasad, P.N., Singh, A.J., Ranjit, A., Narayanan, L.M. and Natarajan, C.R. 1996. Ethnobotany of the Kanikkars of South Tamil Nadu-I: 292-298. In Maheshwari, J.K. (**ed.**) *Ethnobotany in South- Asia.* Scientific Publishers, Jodhpur, India.

Pullaiah, T., Kumar, T. Dharma Chandra. 1996. Herbal plants in Mannanur forest Mahboobnagar Distt., A.P: 218-220. In Maheshwari, J.K. (**ed.**) *Ethnobotany in South Asia.* Scientific Publishers, Jodhpur, India.

Punjani, Bhaskar L. 2002. Ethnobotanical aspects of some weeds from Gujarat. *Ethnobotany* **14**: 78-80.

Purohit, A.N. 2000. Medicinal plants- need for upgrading technology for trading the traditions: 49-75. In Natiyal, A.R., Nautiyal, M.C. and Purohit, A.N. (**ed.**) *Harvesting Herbs. Medicinal and Aromatic Plants. An Action Plan For Uttarkhand.* Bishen Singh Mahendra Pal Singh. Dehradun.

Radhakrishnana, R., Pandurangan, A.G. and Pushpangadan, P.1996. Less known medicinal plants of Kerala state and their conservation. *Ethnobotany* **8**: 82-84.

Raghupathy, S. and Mahadevan, A. 1991. Ethnobotany of Kodiakkarai reserve forest *Ethnobotany* **3**: 79-82.

Rajan, G. Bhaskar, Chezhiyan, N., Khan, Irfan Ali and Khanum, Atiya 2002. Role of medicinal plants for heart and respiratory sytems.: 174-206. In Khan, Irfan Ali and Khanum, Atiya (**ed.**) *Role of Biotechnomolgy in Medicinal and Aromatic Plants.* Ukaaz Publications, Hyderabad, A.P, India.

Rajan, S., Gupta, H.C. and Kumar, Sunil 1999. Exotic medicinal plants of Lucknow. *J. Econ. Taxon. Bot.* **23**: 205-222.

Rajasekaran, S., Pushpagandan, P. and Biju, S.D. 1996. Folk medicine of Kerala: A study on native traditional folk healing art and its practitioners.: 167-172. In Jain, S.K. (**ed.**) *Ethnobiology in Human Welfare.* Deep Publications.

Rajendran, S.M. and Mehrotra, B.N. 1996. Unrecorded medicinal uses of plants among Parambikulam tribals, Kerala, India: 181-183. In Jain, S.K. (**ed.**) *Ethnobiology In Human Welfare.* Deep Publications, New Delhi.

Rajendran, S.M., Sekar, K.Chandra and Sundaresan, V. 2001. Ethnomedicinal lore of Seithur Hills-southern Western Ghats, Tamil Nadu. *Ethnobotany* **13**:101-109.

Raju, R. A. 2000. *Wild Plants of Indian Sub-Continent and Their Economic Use.*CBS Publishers and Distributors, New Delhi-110002.

Rajwar, G.S. 1983. Low altitude medicinal plants of south Garhwal. *Bull. Medico-ethnobot. Res.* **4**: 14-28.

Rana, T.S., Datt, Bhaskar. and Rao, R.R.1996. Strategies for sustainable utilisation of plant resources by the tribals of the Tons valley, Western Himalaya. *Ethnobotany.* **8:** 96-104.

Ranjan, Vinay 1996. Some ethnomedicinal plants of Lalitpur District, Uttar Pradesh, India: 149-150. In Jain, S.K. (**ed**.) *Ethnobiology In Human Welfare.* Deep Publication, New Delhi.

Rao, D. Muralidhar and Pullaiah. T. 2001. Ethno-medico-botanical studies in Guntur Distt. of A.P, India. *Ethnobotany* **13**: 40-44.

Rao, M.K. Vasudeva and Shanpru, R. 1981. Some plants in the life of Garos of Meghalaya.: 153-160. In Jain, S.K. (ed.): *Glimpses of Indian Ethnobotany.* Oxford and IBH Publishing Co., New Delhi.

Rao, N. Rama and Henry, A.N. 1996. *The Ethnobotany of Eastern Ghats in Andhra Pradesh, India.* Botanical Survey of India.Calcutta.

Rao, R.R.1981.Ethnobtanical studies on the flora of Meghalaya–Some interesting reports of herbal medicine:137-148. In Jain, S.K. (ed.) *Glimpses of Indian Ethnobotany.*Oxford and IBH Publishing Co., New Delhi.

Rao, R.R. 1996. Traditional knowledge and sustainable development: Key role of ethnobiologists. *Ethnobotany* **8**:14-24.

Rao, R.R. and Jamir, N.S. 1982. Ethnobotanical studies in Nagaland–II: 54 Medicinal plants used by Nagas. *J. Econ. Tax. Bot.* **3**:11-13.

Rao, R.R. and Neogi.1980. Observations on the ethnobotany of the Khasi and Garo tribes in Meghalaya. *J. Econ. Tax. Bot.* **1**: 163-168.

Rastogi, R.P. and Mehrotra, B. N.1990. *Compendium of Indian Medicinal Plants.*Vol. I. 1960-1969. CDRI, Lucknow. Publications and Information Directorate, New Delhi.

Rastogi, R.P. and Mehrotra, B. N. 1991. *Compendium of Indian Medicinal Plants.* Vol. II 1970-1979. CDRI, Lucknow. Publications and Information Directorate, New Delhi.

Rastogi,R.P. and Mehrotra, B.N. 1993. *Compendium of Indian Medicinal Plants.* Vol. III 1980-1984. CDRI, Lucknow. Publications Information Directorate, New Delhi.

Rastogi, R. P. and Mehrotra, B. N.1995. *Compendium of Indian Medicinal Plants.* Vol. IV. 1985-1989. CDRI, Lucknow. Publications Information Directorate, New Delhi.

Rastogi, R.P. and Mehrotra, B.N. 1995. *Compendium of Indian Medicinal Plants.* Vol. V 1990-1994. CDRI, Lucknow. Publications Information Directorate, New Delhi.

Rawat, G.S. and Pangtey, Y.P.S. 1987. A contribution to the ethnobotany of Alpine regions of Kumaon. *J. Econ. Tax. Bot.* **11**: 139-148.

Roxburgh, W. 1932. Today and Tomorrow Printers and Publications, New Delhi

Roy, G.P. and Chaturvedi, K.K. 1986. Ethnomedicinal traces of Abujh- Marh area, M.P. *Folklore* **27**: 95-100.

Saini, D.C. 1996. Ethnobotany of Thorus of Basti Distt., U.P.: 138-153. In Maheshwari, J.K. (**ed.**) *Ethnobotany in South Asia*. Scientific Publishers, Jodhpur, India.

Saklani, S. and Jain, S.K. 1994. *Cross- Cultural Ethnobotany of North- East India*. Deep Publication, New Delhi.

Sastry, M.S.1961. Comparative chemical study of two varieties of Galangal. *Indian J. Pharm.* **23**: 76.

Satapathy, K.B. and Brahmam, M. 1996. Some medicinal plants used by tribals of Sundergarh District.: 153-158. Jain, S.K. (**ed**.) *Ethnobiology In Human Welfare.* Deep Publication, New Delhi.

Satyavati, G.V., Raina,M.K. and Sharma, M. 1976. *Medicinal Plants of India*. ICMR, New Delhi.

Satyavati, G.V., Tandon, N. and Sharma, M. 1989. Indigenous plant drugs for diabetes mellitus. *Diabetes Bulletein* 181.

Saxena, A.P. and Vyas, K.M. 1981. Ethnobotanical records on infectious diseases from tribals of Vanda district. *J. Econ. Tax. Bot.* **2**: 191-194.

Saxena, A.P. and Vyas, K.M. 1983. Ethnobotany of Dhasan Valley. *J. Econ. Tax. Bot.* **4**:121-128.

Saxena, H.O. and Dutta, P.K. 1975. Studies on the ethnobotany of Orissa. *Bull. Bot Surv. India.* **17**:124-131.

Saxena, H.O., Brahman, M. & Dutta, P.K. 1981. Ethnobotanical studies in Orissa, 232-244. *In:* Jain, S.K. (**ed**.) *Glimpses of Indian Ethnobotany*. Oxford & IBH, New Delhi.

Seetharam, Y.N., Chalageri, Gururaj. Haleshi.C. and Vijay. 1999. Folk medicine of north eastern Karnataka. *Ethnobotany* **11**:32-37

Sen, Soma and Batra, Amla. 1997. Ethno-medico- botany of household remedies of Phagi Tehsil of Jaipur (Rajasthan). *Ethnobotany* **9**: 122-128.

Shah, G.L.1984. Some economically important plants of Salsette island near Bombay. *J. Econ. Tax. Bot.* **5**: 753-765.

Shah, G.L. and Gopal, G. V. 1986. Folklore medicine of Vasavas, Gujarat, India. *Acta Bot. Indica* **14**: 48-53.

Shah, G.L., Menon, A.R. and Gopal, G.V.1981. An account of the ethnobotany of Saurashtra in Gujarat State (India). *J. Econ. Tax. Bot.* **2**: 173-182.

Shah, G.L., Yadav, S.S. and Badri, N. 1983. Medicinal plants from the Dhanau forest division in Maharashtra state. *J. Econ. Tax. Bot.* **4**:141-151.

Shah, N.C. 1982. Herbal folk medicines in northern India. *J. Ethnopharmacol* **6**: 293-301.

Shah, N.C. and Joshi, M.C. 1971. An ethnobotanical study of Kumaon region of India. *Econ. Bot.* **25**: 414-422.

Shah, N.C. and Singh. S.C.1990. Hitherto unreported phytotherapeutical uses from tribal pockets of M.P., India. *Ethnobotany* **2**: 91-95.

Sharma, B.D.and Rana, J.C.1999. Traditional medicinal uses of plants of Himachal Hills. *J. Econ. Taxon. Bot.* **27**:173-176.

Sharma, B.D. and Vyas, M.S. 1985. Ethnobotanical studies on the ferns and fern allies of Rajasthan. *Bull. Bot. Surv. India* **27**: 90-91.

Sharma, K. R. and Sood, Meenu 1997. *Important Medicinal Plants of Himachal Pradesh: I.* Directorate of Extension Education. Dr. Y. S. Parmar University of Horticulture And Forestry, Nauni-Solan (H.P).

Sharma, P.C., Yelne, M.B. and Dennis, T.J. 2001. Data Base on Medicinal Plants Used in Ayurveda. Vol. 1-II. C.C.R.A.S., New Delhi.

Sharma, P.P. 1999. Less known ethnobotanical use of plants in Dadra and Nagar Haveli (U.T): Part I (A-K). *Ethnobotany* **11**: 109-114.

Sharma, P.P. & Singh, N.P. 2001. Ethnomedicinal uses of some edible plants in Dadra Nagar Haveli & Daman (U.T.). *Ethnobotany* **13**: 121-125.

Sharma, Ravindra. 2003. *Medicinal Plants of India –An Encyclopaedia.*Daya Publishing House, Delhi.

Sharma, P.K., Dhyani, S.K. and Shanker, V.1979. Some useful and medicinal plants of the District Dehradun and Siwalik. *J. Sci. Res. Pl. Med.* **1**: 17-43.

Sharma, U.K. 2004. *Medicinal Plants of Assam.* Bishen Singh Mahendra Pal Singh, Dehra Dun.

Shekhawat, G.S. & Anand, S. 1984. An ethnobotanical profile of Indian Desert. *J. Econ. Tax. Bot.* 5: 591-598.

Shome. Usha., Rawat, A.K.S. and Mehrotra, Shanta. 1996. Time tested household herbal remedies.: 96-100. In Jain, S.K. (**ed**.) *Ethnobiology in Human Welfare.* Deep Publications, New Delhi.

Shukla, Gyanesh and Verma, B.K. 1996. Roots–A vital plant part to cure body ailments among tribals/ rural folklore of Western Bihar: 392- 394. In Maheshwari, J.K. (**ed**.) *Ethnobotany in South Asia.* Scientific Publishers, Jodhpur, India.

Singh, Ajai Kumar.1999. A contribution to ethnobotany of sub-Himalayan region of eastern Uttar Pardesh. *J. Econ. Taxon. Bot.* **23**: 237-246.

Singh, Brahma, Chaurasia, O.P. and Jadhav, K.L.1996.An ethnobotanical study of Indus Valley (Ladakh): 92-101. In Maheshwari, J.K. (**ed**.) *Ethnobotany in South Asia.* Scientific Publishers, Jodhpur, India.

Singh, G.C. 2002. Minor forest products of Sariska national park: an ethnobotanical profile: 221-238. In Trivedi, P.C. (**ed**.). *Ethnobotany.* Aavishkar Publishers & Distributors. Jaipur (Raj.), India.

Singh, G.S. 2003. Ethnobotanical inventory of Sariska national park in the Aravali Hills of Eastern Rajasthan, India: 181-195. In Singh, V. & Jain, A.P. (**ed**.). Vol. ². *Ethnobotany and Medicinal Plants of India and Nepal.* Scientific Publishers, Jodhpur, India.

Singh, Gopal S.1999. Ethnobotanical study of the useful plants of Kullu District in north western Himalaya, India. *J. Econ. Taxon. Bot.* **23**: 185-198.

Singh, H.B., Hynneiwta, T.M. & Bora, P. J. 1997. Ethno-medico-botanical studies in Tripura, India. *Ethnobotany* **9**: 56-58.

Singh, K.K. 1996. Ethnobotanical observation on Sonbhadra Distt. of southern U.P, India- utilisation and conservation: 145-148. In Jain, S.K. (**ed**.) *Ethnobiology in Human Welfare*. Deep Publication, New Delhi.

Singh, K.K. and Kumar, Kaushal 2000. *Ethnobotanical Wisdom of Gaddi Tribe in Western Himalaya*. Bishen Singh Mahendra Pal Singh.Dehradun.

Singh, K.K. and Maheshwari, J.K. 1983. Traditional phytotherapy amongst the tribals of Varanasi Distt., U.P. *J. Econ. Tax. Bot.* **4**: 829-838.

Singh, K.K. and Prakash, Anand 1994. Indigenous phytotherapy among the Gond tribe of U.P, India. *Ethnobotany.* **6**: 37-41.

Singh, K.K.and Prakash, Anand 1996. Observation on ethnobotany of the Kol Tribe of Varanasi Distt., U.P., India: 133-137. In Maheshwari, J.K. (**ed**.) *Ethnobotany in South Asia*. Scientific Publishers, Jodhpur, India.

Singh, P.B.1999. *Illustrated Field Guide to Commercially Important Medicinal and Aromatic Plants of Himachal Pradesh (With Special reference to Mandi District)*. Society for Herbal Medicine and Himalayan Biodiversity, Mandi, Himachal Pradesh.

Singh, S.P., Tripathi, S. & Shukla, R.S. 2003. Ethnomedicinal heritage for bioprospecting and drug development in North Eastern States of India: 384-395. In Singh, V. & Jain, A.P. (**ed**.). Vol.-². *Ethnobotany and Medicinal Plants of India and Nepal*. Scientific Publishers, Jodhpur, India.

Singh, Virendra 1996. Ethnomedicobotany of Dards tribe of Gurez valley in Kashmir Himalaya: 129-132. In Jain, S.K. (**ed**.) *Ethnobiology in Human Welfare*. Deep Publication, New Delhi.

Singh, V.P.1973. Some medicinal ferns of Sikkim Himalaya. *J. Res. Indian Med.* **8**: 71-73.

Singh, V. and Pandey, R.P.1980. Medicinal plantlore of the tribals of eastern Rajasthan. *J. Econ. Tax. Bot.* **1**:137-147.

Singh, V. and Pandey, R.P.1998. *Ethnobotany of Rajasthan, India*. Scientific Publishers, India.

Singh, W., Wadhwani, A.M. and Johri, B.M. 1983. *Dictionary of Economic Plants of India (Revised)*. I.C.A.R., New Delhi.

Sinha, B.K. 1996. Some promising medicinal plants among the tribes of Bay islands in Bay of Bengal: 159-161. In Jain, S.K. (**ed**.) *Ethnobiology in Human Welfare*. Deep Pulication, New Delhi.

Sinha, B.K. and Dixit, R.D. 2003. Ethnomedicinal plants sold in Omkareshwar, M.P. *Ethnobotany* **15**: 127-128.

Sinha, B.K., Srivastava, S.K. and Dixit, R.D. 2002. Potential economic legumes of Chhattisgarh and M.P States, India. *J. Econ. Taxon. Bot.* **26**: 587-596.

Sinha, Rajiv K. and Sinha, Shweta. 2001. *Ethnobiology (Role of Indigenous and Ethnic Societies in Biodiversity Conservation, Human Health Protection and Sustainable Development).* Surabhi Publications, Jaipur.

Sinha, V. and Pandey, R.P. 1980. Medicinal plantlore of the tribals of eastern Rajasthan. *J. Econ. Tax. Bot.* **1**:137-147.

Smith, H.F. 2000. The Universal interest of traditional medicine: 1-11. In Natiyal, A.R., Nautiyal, M.C. and Purohit, A.N. (**ed**) *Harvesting Herbs-2000. Medicinal and Aromatic Plants- An Action Plan for Uttarakhand.* Bishen Singh Mahendra Pal Singh, Dehradun.

Sood, S.K., Nath, Ram and Kalia, D.C. 2001. *Ethnobotany of Cold Desert Tries of Lahoul Spiti.* Deep Publications, New Delhi.

Subramaniam,A. and Babu, V. 2003. Standarised phytomedicines for diabetes: 46-62. In Khan, Irfan Ali and Khanum, Atiya (**ed.**) *Role of Biotechnomolgy in Medicinal and Aromatic Plants.* Ukaaz Publications, Hyderabad, A.P, India.

Sudhakar, S. and Rolla, R.S. 1985. Medicinal plants of upper east Godavari Distt., A.P. and need for establishment of medicinal farm. *J. Econ. Tax. Bot.* **7**:399- 406.

Sur, P.K. 2002. Taxonomy and ethnobotany of genus *Mucuna adans* in India. *J. Econ. Taxon. Bot.* **26**: 494-498.

Swami, Ajai and Gupta, B.K. 1996. A note on some commonly occuring medicinal weeds of Udhampur Distt. (J.K): 89-91. In Maheshwari, J.K. (**ed.**) *Ethnobotany in South Asia.* Scientific Publishers, Jodhpur, India.

Tarafdar, C.R. 1983a. Ethnogynaecology in relation to plants I- Plants used in antifertility and conception. *J. Econ. Tax. Bot.* **4**: 483-489.

Tarafdar, C.R. 1983b. Ethnogynaecology in relation to plants II- Plants used for abortion. *J.Econ. Tax. Bot.* **4**: 507-516.

Tarafdar, C.R. 1984. Ethnogynaecology in relation to plants II- Plants used to accelerate delivery and in pre- and post-natal complaints. *J. Econ. Tax. Bot.* **5**: 572-576.

Tarafdar, C.R. 1987. Some traditional knowledge about tribal health. *Folklore* **28**: 37-42.

Tarafdar, C.R.1986. Ethnobotany of chhota Nagpur (Bihar). *Folklore* **27**: 119-124.

Tarafdar, C.R. and Chaudhary, H.N. R. 1981. Less known medicinal uses of plants among the tribals of Hazaribagh Distt. of Bihar: 208-217. In Jain, S.K. (ed.) *Glimpses of Indian Ethnobotany.* Oxford and IBH Publishing Co., New Delhi.

Thakur, R.S., Puri, H.S. and Hussain, Akhtar 1989. *Major Medicinal Plants of India.* Central Institute of Medicinal and Aromatic Plants. Lucknow. India.

Tiwari, J.K. 1999. Ethnobotanical notes on the Flora of Har-Ki- Doon (Disst. Uttarkhand) Garhwal Himalaya, U.P, India. *Ethnobotany* **11**: 9-17.

Tripathi, G. and Kumar, A. 2003. *Potentials of Living Resources.* Discovery Publishing House, New Delhi.

Tripathy, Sunil and Goel, Anil 2001. Ethnobotanial diversity of zingiberaceae in north-eastern India. *Ethnobotny* **13**: 67-79.

Trivedi, Praveen Chandra. 2002. *Ethnobotany*. Aavishkar Publishers, Distributers, Jaipur 302003 (Raj.) India.

Tosh, Jayananda 1996. Ethnobotanical study of western Maharashtra.: 169-174. In Maheshwari, J.K. (**ed**.) *Ethnobotany in South-Asia*. Scientific Publishers, Jodhpur, India.

Uniyal, B.P. and Malhotra, C.L.1989. Economic exploitation of rare north-western Himalaya plants. In Paliwal, G.S. (**ed**.) *The Vegetational Wealth of the Himalyas*.Puja Publishers.Delhi.

Upadhyay, R. and Chauhan, S.V.S. 2000. Ethnobotanical observations on Koya tribe of Gundala mandal of Khamman district, Andhra Pradesh. *Ethnobotany* **12**: 93-99.

Upadhye, Anuradha S., Vartak, V.D. and Kumbhojkar, M.S. 1994. Ethno-medico-botanical studies in western Maharashtra, India. *Ethnobotany* **6**: 25-31.

Varma, S.K. 1997. Comparative studies on folk drugs of tribals of chota Nagpur and Santhal Pargana of Bihar, India. *Ethnobotany* **9**: 70-76.

Verma, P., Khan, A.A. and Singh, K.K.1995. Traditional phytotherapy among the Baigu Tribe of Shahdol District of Madhya Pradesh, India. *Ethnobotany.***7**: 69-73.

Vieira, L.S. 1992. *Fitoterapia de Amazonia: Manual de plantas Medicinais Sao Paulo*. Agronomica Ceres.

Vir Jee, Dar, G.H., Kachroo, P. and Bhatt, G.M. 1984. Taxo- ethnobotanical studies of the rural areas in Distt. Rajouri (Jammu). *J. Econ. Tax. Bot.* **5**: 831-838.

Viswanathan, M.V. 1999. Edible and medicinal plants of Ladakh (J&K). *J. Econ. Tax. Bot. 23* (1): 151-154.

Vishwanathan, M.B., Kumar, Prem. E.H. and Ramesh, N. 2001. Ethnomedicines of Kanis in Kalakkad- Mundanthurai tiger reserve, T.N. *Ethnobotany* **13**: 60-66.

Viswanathan, M.V. and Singh, H.B. 1996. Plants used as household remedies in India: 105-107. In Jain, S.K. (**ed**.) *Ethnobiology In Human Welfare*. Deep Publications, New Delhi.

Wambebe, C.O.N. 1990. National products in developing economy. In. Igbocchi, A.C. and Osisigu, I.C.W. (**ed**.) *National Workshop on Natural Products*. University of Benin press, Nigeria.

Warrier, P.K., Nambiar, V.P.K. and Kutty, C.R. 1994-96. *Indian Medicinal Plants*. Vol. 1-5. Orient Longman Ltd., Madras.

Watt, G. 1972. *A Dictionary of the Economic Products of India*. Vol.1-6. Peridicals Experts, New Delhi.

Yusuf, Mohammad., Chowdhary, M.A., Rahman, Jashin Uddin and Begum, Jaripa. 2002. Indigenous knowledge about the use of zingiber in Bangladesh. *Journal of Econ. Taxon. Bot.*: **26**: 566-570.

Appendices

Appendix–I
Index to Total Number of Genera and Species Under Various Divisions of Plant Kingdom for Cure of Cold, Flu and Lungs Ailments

	Total No. of Genera and Species	*Generic Name and Total Number of Species Used for the Cure of Lungs Ailments*
FUNGI	(2/2)	*Agaricus* (1), *Morchella* (1).
LICHEN	(1/1)	*Usnea* (1).
PTERIDOPHYTES	(12/16)	*Actiniopteris* (1), *Adiantum* (5), *Chelianthes* (1), *Dicranopteris* (1), *Diplazium* (1), *Gleichenia* (1), *Lycopodium* (1), *Marsilea* (1), *Nephrolepis* (1), *Ophioglossum* (1), *Pellalaeo*, (1), *Tectaria* (1).
GYMNOSPERMS	(8/12)	*Abies* (1), *Cupressus* (1), *Ephedra* (1), *Ginkgo* (1), *Juniperus* (3), *Pinus* (3), *Taxodium* (1), *Taxus* (1).
MONOCOTS	(47/60)	*Acorus* (1), *Allium* (2), *Aloe* (1), *Alpinia* (3), *Amomum* (1), *Amorphophallus* (1), *Aneilema* (1), *Aplectrum* (1), *Areca* (1), *Arisaema* (1), *Bambusa* (2), *Borassus* (1), *Chlorophytum* (1), *Cocos* (1), *Costus* (1), *Crinum* (1), *Crocus* (1), *Curculigo* (1), *Curcuma* (4), *Cymbopogon* (1), *Cyperus* (1), *Dioscorea* (1), *Fritillaria* (2), *Globba* (2), *Gloriosa* (1), *Hedychium* (2), *Hordeum* (1), *Hornstedtia* (1), *Kaempferia* (1), *Kyllinga* (1), *Monochoria* (1), *Musa* (1), *Pandanus* (1), *Paris* (1), *Phoenix* (2), *Pothos* (1), *Rhynchanthus* (1), *Sansevieria* (1), *Sauromatum* (1), *Scindapsus* (1), *Serenoa* (1), *Symplocarpus* (1), *Tacca* (1), *Urginea* (1), *Vanda* (2), *Zea* (1), *Zingiber* (2).
DICOTS	(328/439)	*Abelmoschus* (2), *Abrus* (1), *Abutilon* (1), *Acacia* (4), *Acalypha* (1), *Acanthus* (1), *Achillea* (1), *Achyranthes* (1), *Aconitum* (1), *Adansonia* (1), *Adhatoda* (1), *Aerva* (1), *Ageratum* (1), *Ailanthus* (3), *Ajuga* (1), *Alangium* (1), *Albizia* (2), *Alhagi* (1), *Alstonia* (1), *Althaea* (1), *Alysicarpus* (3), *Amaranthus* (1), *Anacardium* (1), *Andrographis* (1), *Angelica* (2), *Anisochilus* (1), *Anogeissus* (1), *Apium* (1), *Apocynum* (1), *Arctium* (1), *Argemone* (1), *Aristolochia* (2), *Arnebia* (1), *Artemisia* (3), *Asclepias* (1), *Asclepiodora* (1), *Aster* (1), *Astragalus* (1), *Atropa* (2), *Atylosia* (1), *Azadirachta* (1), *Azima* (1), *Bacopa* (1), *Balanites* (1), *Balanophora* (1), *Baliospermum* (1), *Barleria* (1), *Barringtonia* (1), *Bauhinia* (1), *Benincasa* (1), *Bergenia* (1), *Betonica* (1), *Betula* (1), *Biophytum* (1), *Blumea* (1), *Boerhaavia* (1), *Bombax* (1), *Boswellia*

Contd..

Appendix–I–Contd...

Total No. of Genera and Species	*Generic Name and Total Number of Species Used for the Cure of Lungs Ailments*
	(1), *Brassica* (2), *Brophytum* (1), *Butea* (1), *Byttneria* (1), *Caesalpinia* (2), *Cajanus* (1), *Calotropis* (2), *Cannabis* (1), *Capparis* (2), *Capsicum* (1), *Carica* (1), *Carissa* (1), *Carmona* (1), *Casearia* (1), *Cassia* (7), *Castanea* (1), *Castanopsis* (1), *Catha* (1), *Celastrus* (1), *Celosia* (1), *Centella* (1), *Cephalanthus* (1), *Cephaelis* (1), *Chasalia* (1), *Cicer* (1), *Cimicifuga* (1), *Cinnamomum* (4), *Cissampelos* (1), *Cissus* (1), *Citrullus* (1), *Citrus* (1), *Clematis* (1), *Clerodendrum* (2), *Clitoria* (1), *Coffea* (1), *Colebrooka* (1), *Coleus* (1), *Commiphora* (1), *Corchorus* (1), *Cordia* (2), *Croton* (1), *Cuscuta* (1), *Cydonia* (1), *Dalbergia* (1), *Datura* (3), *Delphinuumhs* (1), *Dendropthoe* (1), *Descuraina* (1), *Desmodium* (3), *Desmos* (1), *Digitalis* (1), *Dillenia* (1), *Diospyros* (2), *Drosera* (1), *Drymaria* (1), *Eclipta* (1), *Elaeagnus* (2), *Elephantopus* (1), *Elsholtzia* (1), *Elytraria* (1), *Embelia* (1), *Emblica* (1), *Epimedium* (1), *Erioglossum* (1), *Erycibe* (1), *Erythrina* (1), *Erythrophloem* (1), *Erythroxylon* (1), *Eucalyptus* (1), *Eupatorium* (1), *Euphorbia* (4), *Evolvulus* (1), *Fagonia* (2), *Ferula* (2), *Ficus* (3), *Foeniculum* (1), *Galeopsis* (1), *Garuga* (1), *Gentianella* (1), *Geranium* (1), *Glycyrrhiza* (1), *Gymnema* (1), *Gynandropsis* (1), *Hedyotis* (1), *Helianthus* (1), *Helichrysum* (1), *Helicteres* (1), *Hemidesmus* (1), *Hepatica* (1), *Heracleum* (1), *Hibiscus* (1), *Hippophae* (1), *Hiptage* (1), *Holarrhena* (1), *Holoptelea* (1), *Holostemma* (1), *Humboldtia* (1), *Hymenostegia* (1), *Hyoscyamus* (1), *Hypericum* (1), *Hyssopus* (1), *Iberlis* (1), *Indigofera* (3), *Inula* (1), *Jasminum* (2), *Jatropha* (1), *Justicia* (1), *Kalanchoe* (2), *Kedrostis* (1), *Lactuca* (2), *Laggera* (1), *Lannea* (1), *Lantana* (1), *Leea* (2), *Lens* (1), *Leonurus* (1), *Lepidium* (1), *Leptadenia* (1), *Leucas* (5), *Lindenbergia* (1), *Lindera* (1), *Linum* (1), *Lippia* (1), *Litsea* (1), *Lobelia* (3), *Lonicera* (1), *Luffa* (2), *Machillus* (1), *Madhuca* (1), *Maerua* (1), *Malva* (4), *Mangifera* (1), *Marrubium* (1), *Melaleuca* (1), *Melilotus* (1), *Melissa* (1), *Mentha* (2), *Mesua* (1), *Michelia* (1), *Milletia* (1), *Mimosa* (1), *Momordica* (1), *Monarda* (1), *Morinda* (1), *Moringa* (1), *Mucuna* (1), *Murraya* (1), *Mussaenda* (2), *Myrica* (2), *Myroxylon* (1), *Myrtus* (1), *Nardostachys* (1), *Naregamia* (1), *Nelumbo* (1), *Nepeta* (1), *Nigella* (1),

Contd..

Appendix–I–Contd...

Total No. of Genera and Species	Generic Name and Total Number of Species Used for the Cure of Lungs Ailments
	Ochna (2), *Ocimum* (3), *Oldenlandia* (2), *Oligochaeta* (1), *Opuntia* (2), *Origanum* (1), *Osbeckia* (1), *Oxalis* (1), *Paederia* (1), *Passiflora* (1), *Pedalium* (1), *Peganium* (1), *Pergularia* (1), *Pericampylus* (1), *Phaseolus* (1), *Phlogacanthus* (1), *Phyllanthus* (2), *Picrorrhiza* (1), *Pillostigma* (1), *Pimpinella* (1), *Piper* (4), *Pistacia* (2), *Pittosporum* (1), *Plantago* (2), *Plectranthus* (1), *Podophyllum* (1), *Pogostemon* (1), *Polyalthia* (1), *Polygala* (4), *Polygonum* (2), *Pongamia* (1), *Populus* (1), *Portulaca* (2), *Potentila* (1), *Pratia* (1), *Premna* (1), *Prosopis* (1), *Prunus* (4), *Pseudarthria* (1), *Psidium* (1), *Pterocarpus* (2), *Pterospermum* (1), *Punica* (1), *Quercus* (1), *Ranunculus* (1), *Raphanus* (1), *Rheum* (1), *Rhodiola* (1), *Rhododendron* (1), *Rhus* (1), *Ricinus* (1), *Rorippa* (2), *Rosmarinus* (1), *Rotala* (1), *Rumex* (1), *Ruta* (1), *Salacia* (1), *Salix* (1), *Salvadora* (1), *Salvia* (2), *Sambucus* (1), *Sapindus* (1), *Saussurea* (3), *Scoparia* (1), *Scutellaria* (1), *Selinium* (1), *Semecarpus* (1), *Sesamum* (1), *Sida* (4), *Sisymbrium* (1), *Solanum* (8), *Sonchus* (1), *Sorbus* (1), *Sphaeranthus* (1), *Stellaria* (1), *Stephania* (1), *Stercularia* (1), *Stereospermum* (2), *Strobilanthus* (1), *Strychnos* (2), *Styrax* (1), *Syzygium* (3), *Tagetes* (1), *Tamarindus* (1), *Taraxacum* (1), *Tectona* (1), *Tephrosia* (2), *Terminalia* (3), *Thevetia* (1), *Thymus* (1), *Tinospora* (2), *Toona* (1), *Trachyspermum* (1), *Tragia* (1), *Tragopogon* (1), *Trema* (2), *Trianthema* (1), *Tribulus* (1), *Trigonella* (1), *Tussilago* (1), *Tylophora* (1), *Vateria* (1), *Verbascum* (1), *Vernonia* (1), *Veronica* (1), *Viola* (6), *Vitex* (1), *Vitis* (1), *Wagatea* (1), *Withania* (1), *Woodfordia* (1), *Wrightia* (1), *Zanthoxylum* (2), *Zizyphus* (2).

Name of Family	Total No. of Genera and Species	Total No. of Species in a Genus
Acanthaceae	(8/8)	*Acanthus* (1), *Adhatoda* (1), *Andrographis* (1), *Barleria* (1), *Elytraria* (1), *Justicia* (1), *Phlogacanthus* (1), *Strobilanthus* (1).
Actiniopteridaceae	(1/1)	*Actiniopteris* (1).
Agavaceae	(1/1)	*Sansevieria* (1).
Agaricaceae	(1/1)	*Agaricus* (1).
Aizoaceae	(1/1)	*Trianthema* (1).
Alangiaceae	(1/1)	*Alangium* (1).
Amaranthaceae	(4/4)	*Achyranthes* (1), *Aerva* (1), *Amaranthus* (1), *Celosia* (1).
Amaryllidaceae	(1/1)	*Crinum* (1).
Anacardiaceae	(6/7)	*Anacardium* (1), *Lannea* (1), *Mangifera* (1), *Pistacia* (2), *Rhus* (1), *Semecarpus* (1).
Annonaceae	(2/2)	*Desmos* (1), *Polyalthia* (1).
Apiaceae	(9/11)	*Angelica* (2), *Apium* (1), *Centella* (1), *Ferula* (2), *Foeniculum* (1), *Heracleum* (1), *Pimpinella* (1), *Selinium* (1), *Trachyspermum* (1).
Apocynaceae	(5/5)	*Alstonia* (1), *Apocynum* (1), *Carissa* (1), *Holarrhena* (1), *Wrightia* (1).
Araceae	(6/6)	*Acorus* (1), *Amorphophallus* (1), *Arisaema* (1), *Sauromatum* (1), *Scindapsus* (1), *Symplocarpus* (1).
Arecaceae	(6/7)	*Areca* (1), *Borassus* (1), *Cocos* (1), *Phoenix* (2), *Pothos* (1), *Serenoa* (1).
Aristolochiaceae	(1/2)	*Aristolochia* (2).
Asclepiadaceae	(9/10)	*Asclepias* (1), *Asclepiodora* (1), *Calotropis* (2), *Gymnema* (1), *Hemidesmus* (1), *Holostemma* (1), *Leptadenia* (1), *Pergularia* (1), *Tylophora* (1).
Aspidiaceae	(1/1)	*Tectaria* (1).
Asteraceae	(24/29)	*Achillea* (1), *Ageratum* (1), *Arctium* (1), *Artemisia* (3), *Aster* (1), *Blumea* (1), *Eclipta* (1), *Elephantopus* (1), *Eupatorium* (1), *Helianthus* (1), *Helichrysum* (1), *Inula*

Contd..

Appendix–II–Contd...

Name of Family	Total No. of Genera and Species	Total No. of Species in a Genus
		(1), *Lactuca* (2), *Laggera* (1), *Oligochaeta* (1), *Saussurea* (3), *Sonchus* (1), *Sphaerantus* (1), *Tagetes* (1), *Taraxacum* (1), *Tragopogon* (1), *Tussilago* (1), *Vernonia* (1), *Veronica* (1).
Balanitaceae	(1/1)	*Balanites* (1).
Betulaceae	(1/1)	*Betula* (1).
Bignoniaceae	(1/2)	*Stereospermum* (2).
Bombacaceae	(2/2)	*Adansonia* (1), *Bombax* (1).
Boraginaceae	(3/4)	*Arnebia* (1), *Carmona* (1), *Cordia* (2).
Brassicaceae	(7/9)	*Brassica* (2), *Descurainia* (1), *Iberis* (1), *Lepidium* (1), *Raphanus* (1), *Rorippa* (2), *Sisymbrium* (1).
Burseraceae	(4/4)	*Boswellia* (1), *Commiophora* (1), *Garuga* (1), *Tragia* (1).
Cactaceae	(1/2)	*Opuntia* (2).
Campanulaceae	(2/4)	*Lobelia* (3), *Pratia* (1).
Cannabinaceae	(1/1)	*Cannabis* (1).
Capparidaceae	(3/4)	*Capparis* (2), *Gynandropsis* (1), *Maerua* (1).
Caprifoliaceae	(2/2)	*Lonicera* (1), *Sambucus* (1).
Caricaceae	(1/1)	*Carica* (1).
Caryophyllaceae	(2/2)	*Drymaria* (1), *Stellaria* (1).
Celastraceae	(2/2)	*Catha* (1), *Celastrus* (1).
Clusiaceae	(1/1)	*Mesua* (1).
Combretaceae	(2/4)	*Anogeissus* (1), *Terminalia* (3).
Commelinaceae	(1/1)	*Aneilema* (1).
Convolvulaceae	(2/2)	*Erycibe* (1), *Evolvulus* (1).
Crassulaceae	(2/3)	*Kalanchoe* (2), *Rhodiola* (1).
Cucurbitaceae	(5/6)	*Benincasa* (1), *Citrullus* (1), *Kedrostis* (1), *Luffa* (2), *Momordica* (1).
Cupressaceae	(2/4)	*Cupressus* (1), *Juniperus* (3).
Cuscutaceae	(1/1)	*Cuscuta* (1).
Cyperaceae	(2/2)	*Cyperus* (1), *Kyllinga* (1).
Davalliaceae	(1/1)	*Nephrolepis* (1).
Dilleniaceae	(1/1)	*Dillenia* (1).

Contd..

Appendix–II–Contd...

Name of Family	Total No. of Genera and Species	Total No. of Species in a Genus
Dioscoreaceae	(1/1)	*Disocorea* (1).
Dipterocarpaceae	(2/2)	*Balanophora* (1), *Vateria* (1).
Droseraceae	(1/1)	*Drosera* (1).
Ebenaceae	(1/2)	*Diospyros* (2).
Elaeagnaceae	(2/3)	*Elaeagnus* (2), *Hippophae* (1).
Ephedraceae	(1/1)	*Ephedra* (1).
Ericaceae	(1/1)	*Rhododendron* (1).
Erythroxylaceae	(1/1)	*Erythroxylon* (1).
Euphorbiaceae	(9/13)	*Acalypha* (1), *Baliospermum* (1), *Barringtonia* (1), *Croton* (1), *Emblica* (1), *Euphorbia* (4), *Jatropha* (1), *Phyllanthus* (2), *Ricinus* (1).
Fabaceae	(38/57)	*Abrus* (1), *Acacia* (4), *Albizzia* (2), *Alhagi* (1), *Alysicarpus* (3), *Astragalus* (1), *Atylosia* (1), *Bauhinia* (1), *Butea* (1), *Caesalpinia* (2), *Cajanus* (1), *Cassia* (7), *Cicer* (1), *Clitoria* (1), *Dalbergia* (1), *Desmodium* (3), *Erythrina* (1), *Erythrophloeum* (1), *Glycyrrhiza* (1), *Humboldtia* (1), *Hymenostegia* (1), *Indigofera* (3), *Lens* (1), *Melilotus* (1), *Milletia* (1), *Mimosa* (1), *Mucuna* (1), *Myroxylon* (1), *Phaseolus* (1), *Piliostigma* (1), *Pongamia* (1), *Prosopis* (1), *Pseudarthia* (1), *Pterocarpus* (2), *Tamarindus* (1), *Tephrosis* (2), *Trigonella* (1), *Wagatea* (1).
Fagaceae	(3/3)	*Castanea* (1), *Castanopsis* (1), *Quercus* (1).
Flacourtiaceae	(1/1)	*Casearia* (1).
Gentianaceae	(2/2)	*Gentianella* (1), *Geranium* (1).
Ginkgoaceae	(1/1)	*Ginkgo* (1).
Gleicheniaceae	(2/2)	*Dicranopteris* (1), *Gleichenia* (1).
Hippocrateaceae	(1/1)	*Salacia* (1).
Hypoxidaceae	(1/1)	*Curculigo* (1).
Hypericaceae	(1/1)	*Hypericum* (1).
Iridaceae	(1/1)	*Crocus* (1).
Lamiaceae	(23/31)	*Ajuga* (1), *Anisochilus* (1), *Betonica* (1), *Colebrooka* (1), *Coleus* (1), *Elsholtzia* (1), *Galeopsis* (1), *Hyssopus* (1), *Leonurus* (1), *Leucas* (5), *Marrubium* (1), *Melissa* (1),

Contd..

Appendix–II–Contd...

Name of Family	Total No. of Genera and Species	Total No. of Species in a Genus
		Mentha (2), *Monarda* (1), *Nepeta* (1), *Ocimum* (3), *Origanum* (1), *Plectranthus* (1), *Pogostemon* (1), *Rosmarinus* (1), *Salvia* (2), *Scutellaria* (1), *Thymus* (1).
Lauraceae	(4/7)	*Cinnamomum* (4), *Lindera* (1), *Litsea* (1), *Machillus* (1).
Leeaceae	(1/2)	*Leea* (2).
Liliaceae	(7/9)	*Allium* (2), *Aloe* (1), *Chlorophytum* (1), *Fritillaria* (2), *Gloriosa* (1), *Paris* (1), *Urginea* (1).
Linaceae	(1/1)	*Linum* (1).
Loganiaceae	(1/2)	*Strychnos* (2).
Loranthaceae	(1/1)	*Dendropthoe* (1).
Lycopodiaceae	(1/1)	*Lycopodium* (1).
Lythraceae	(2/2)	*Rotala* (1), *Woodfordia* (1).
Magnoliaceae	(1/1)	*Michelia* (1).
Malphigiaceae	(1/1)	*Hiptage* (1).
Malvaceae	(7/14)	*Abelmoschus* (2), *Abutilon* (1), *Althaea* (1), *Hibiscus* (1), *Malva* (4), *Sida* (4), *Thvetia* (1).
Marsileaceae	(1/1)	*Marsilea* (1).
Melastomaceae	(1/1)	*Osbeckia* (1).
Meliaceae	(3/3)	*Azadirachta* (1), *Naregamia* (1), *Toona* (1).
Menispermaceae	(4/5)	*Cissampelos* (1), *Pericampylus* (1), *Stephania* (1), *Tinospora* (2).
Moraceae	(1/3)	*Ficus* (3).
Morchellaceae	(1/1)	*Morchella* (1).
Moringaceae	(1/1)	*Moringa* (1).
Musaceae	(1/1)	*Musa* (1).
Myricaceae	(1/2)	*Myrica* (2).
Myrsinaceae	(1/1)	*Embelia* (1).
Myrtaceae	(5/7)	*Eucalyptus* (1), *Melaleuca* (1), *Myrtus* (1), *Psidium* (1), *Syzygium* (3).
Nyctaginaceae	(1/1)	*Boerhaavia* (1).
Nymphaeaceae	(1/1)	*Nelumbo* (1).
Ochnaceae	(1/2)	*Ochna* (2).

Contd..

Appendix–II–Contd...

Name of Family	Total No. of Genera and Species	Total No. of Species in a Genus
Oleaceae	(1/2)	*Jasminum* (2).
Ophioglossaceae	(1/1)	*Ophioglossum* (1).
Orchidaceae	(2/3)	*Aplectrum* (1), *Vanda* (2).
Oxalidaceae	(3/3)	*Biophytum* (1), *Brophytum* (1), *Oxalis* (1).
Pandanaceae	(1/1)	*Pandanus* (1).
Papaveraceae	(1/1)	*Argemone* (1).
Passifloraceae	(1/1)	*Passiflora* (1).
Pedaliaceae	(2/2)	*Pedalium* (1), *Sesamum* (1).
Pinaceae	(2/4)	*Abies* (1), *Pinus* (3).
Pittosporaceae	(1/1)	*Pittosporum* (1).
Piperaceae	(1/4)	*Piper* (4).
Plantaginaceae	(1/2)	*Plantago* (2).
Poaceae	(4/5)	*Bambusa* (2), *Cymbopogon* (1), *Hordeum* (1), *Zea* (1).
Podophyllaceae	(2/2)	*Epimedium* (1), *Podophyllum* (1).
Polygalaceae	(1/4)	*Polygala* (4).
Polygonaceae	(3/4)	*Polygonum* (2), *Rheum* (1), *Rumex* (1).
Polypodiaceae	(2/6)	*Adiantum* (5), *Pellaleo* (1).
Pontederiaceae	(1/1)	*Monochoria* (1).
Pteridaceae	(1/1)	*Diplazium* (1).
Portulacaceae	(1/2)	*Portulaca* (2).
Punicaceae	(1/1)	*Punica* (1).
Ranunculaceae	(7/7)	*Aconitum* (1), *Cimicifuga* (1), *Clematis* (1), *Delphinium* (1), *Hepatica* (1), *Nigella* (1), *Ranunculus* (1).
Rhamnaceae	(1/2)	*Zizyphus* (2).
Rosaceae	(4/7)	*Cydonia* (1), *Potentilla* (1), *Prunus* (4), *Sorbus* (1).
Rubiaceae	(9/11)	*Cephaelis* (1), *Cephalenthus* (1), *Chasalia* (1), *Coffea* (1), *Hedyotis* (1), *Morinda* (1), *Mussaenda* (2), *Oldenlandia* (2), *Paederia* (1).
Rutaceae	(5/6)	*Citrus* (1), *Murraya* (1), *Peganium* (1), *Ruta* (1), *Zanthoxylum* (2).
Salicaceae	(2/2)	*Populus* (1), *Salix* (1).

Contd..

Appendix–II–Contd...

Name of Family	Total No. of Genera and Species	Total No. of Species in a Genus
Salvadoraceae	(2/2)	*Azima* (1), *Salvadora* (1).
Sapindaceae	(2/2)	*Erioglossum* (1), *Sapindus* (1).
Sapotaceae	(1/1)	*Madhuca* (1).
Saxifragaceae	(1/1)	*Bergenia* (1).
Scrophulariaceae	(6/6)	*Bacopa* (1), *Digitalis* (1), *Lindenbergia* (1), *Picrrorhiza* (1), *Scroparia* (1), *Verbascum* (1).
Simaroubaceae	(1/3)	*Ailanthus* (3).
Sinopteridaceae	(1/1)	*Chelianthus* (1).
Solanaceae	(6/16)	*Atropa* (2), *Capsicum* (1), *Datura* (3), *Hyocyamus* (1), *Solanum* (8), *Withania* (1).
Sterculiaceae	(4/4)	*Byttneria* (1), *Helicteres* (1), *Pterospermum* (1), *Stercularia* (1).
Styraceae	(1/1)	*Styrax* (1).
Taccaceae	(1/1)	*Tacca* (1).
Taxaceae	(1/1)	*Taxus* (1).
Taxodiaceae	(1/1)	*Taxodium* (1).
Tilliaceae	(1/1)	*Corchorus* (1).
Ulmaceae	(2/3)	*Holoptelea* (1), *Trema* (2).
Usneaceae	(1/1)	*Usnea* (1).
Valerianaceae	(1/1)	*Nardostachys* (1).
Verbenaceae	(6/7)	*Clerodendrum* (2), *Lantana* (1), *Lippia* (1), *Premna* (1), *Tectona* (1), *Vitex* (1).
Violaceae	(1/6)	*Viola* (6).
Vitaceae	(2/2)	*Cissus* (1), *Vitis* (1).
Zingiberacae	(10/18)	*Alpinia* (3), *Amomum* (1), *Costus* (1), *Curcuma* (4), *Globba* (2), *Hedychium* (2), *Hornstedtia* (1), *Kaempferia* (1), *Rhynchanthus* (1), *Zingiber* (2).
Zygophyllaceae	(2/3)	*Fagonia* (2), *Tribulus* (1).

Appendix–III
Glossary to Some Important Medical Terms

Asthma	:	It is a condition that affects the air passages of the lungs. When a person has asthma, the air passages are inflamed which means that the airways are red and swollen. Inflammation of the air passages makes them over extra-sensitive to a number of different things that can "trigger", or bring on asthma symptoms. Coughing, wheezing (whistling noise during breathing), chest tightness or shortness of breath are one or more of the associated symptoms.
Bronchial Asthma	:	It is an inflammatory disorder of the airways, characterized by periodic attacks of wheezing, shortness of breath, chest tightness and coughing.
Bronchitis	:	Bronchitis is an inflammation of the bronchi- the main air passages to the lungs. Acute bronchitis generally follows a viral respiratory infection. The viral infection produces bronchial inflammation, which sets the stage for bronchitis and secondary bacterial infection. Chronic bronchitis is a long-term condition off excessive bronchial mucus with a productive cough.
Cold	:	A cold is a contagious viral infection of the upper respiratory tract characterized by inflammation of the mucus membranes, sneezing and sore throat.
Expectorant	:	Causing or stimulating expectoration to cough up and spit.
Laryngitis	:	Inflammatory infection of the larynx.
Lungs Diseases	:	Lungs disease is any disease or disorder characterized by impaired functioning of lungs.
Pharyngitis	:	Sore throat or inflammatory disease of the pharynx caused by a number of viruses, bacteria and fungi.
Pthisis	:	A wasting away of the body, or any of its parts, especially in tuberculosis of lungs.
Pneumonia	:	Pneumonia is an inflammation of the lungs caused by a bacterial, viral or fungal infection.
Rubefacient	:	Any external application of the skin.
Tuberculosis	:	Tuberculosis is a contagious bacterial infection caused by *Mycobacterium tuberculosis* (TB). The lungs are primarily involved, but the infection can spread to other organs

Index to Botanical Names

Index to English Names

Index to Hindi Names

Index to Sanskrit Names

Plate 1: Flowers of *Adhatoda vasica* Nees

Plate 2: Flowers of *Allium sativum* Linn.

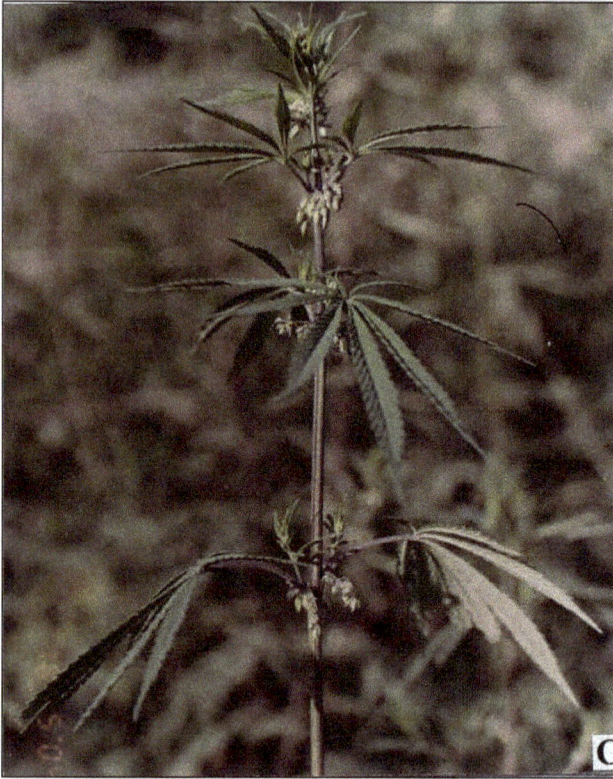

Plate 3: Plant of *Cannabis sativa* Linn.

Plate 4: *Carica papaya* Linn.

Plate 5: Flowering branch of *Cassia* sp.

Plate 6: Fruits of *Emblica officinalis* Gaertn.

Plate 7: *Foeniculum vulgare* Mill.

Plate 8: *Hibiscus rosa-sinensis* Linn.

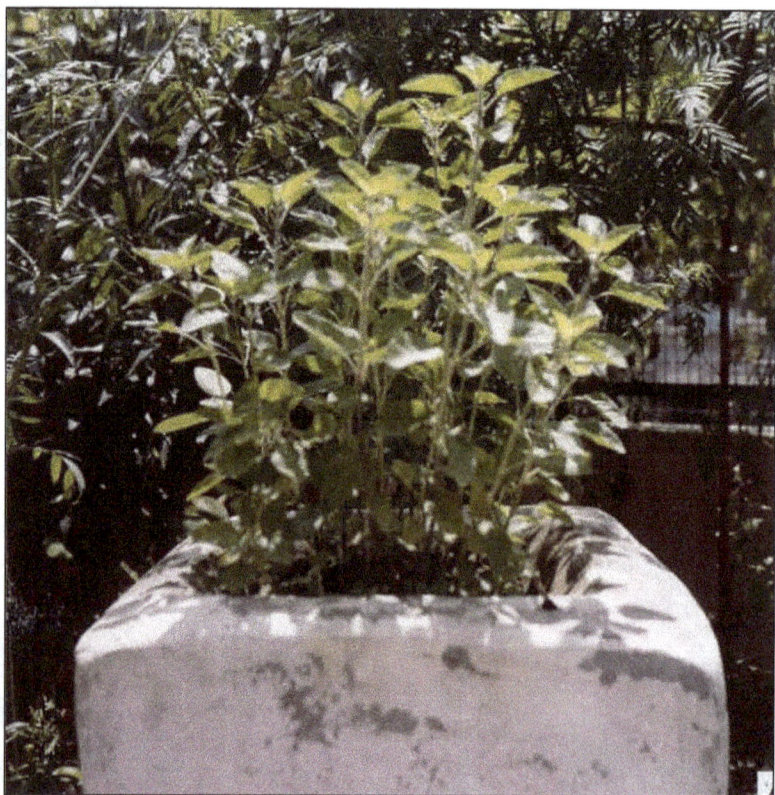

Plate 9: A. *Ocimum sanctum* Linn.

Plate 10: *Iberis amara* **Linn.**

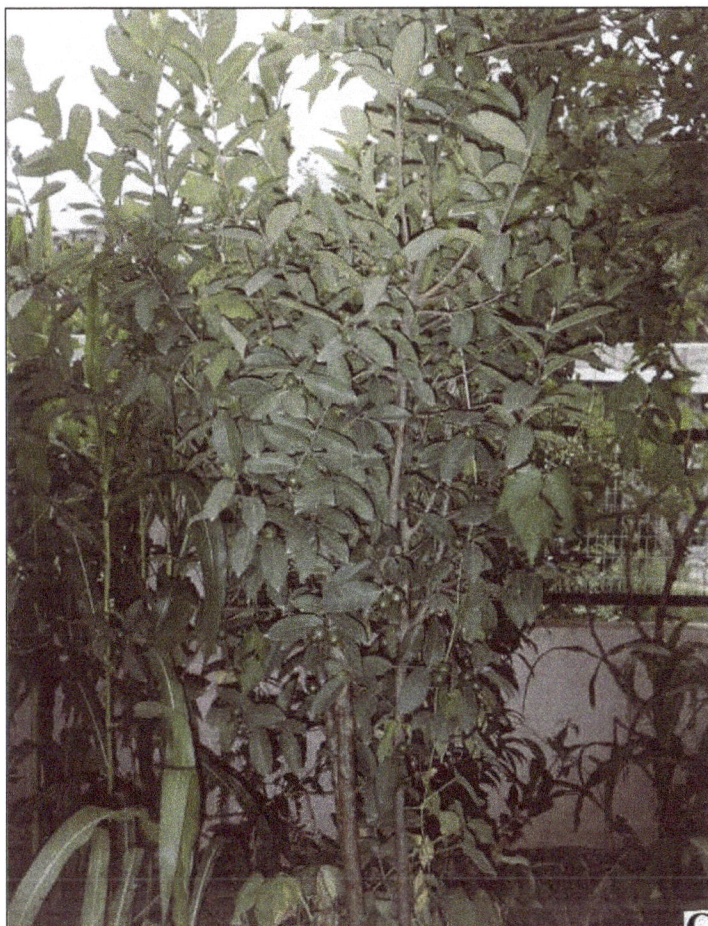

Plate 11: *Psidium guajava* **Linn.**

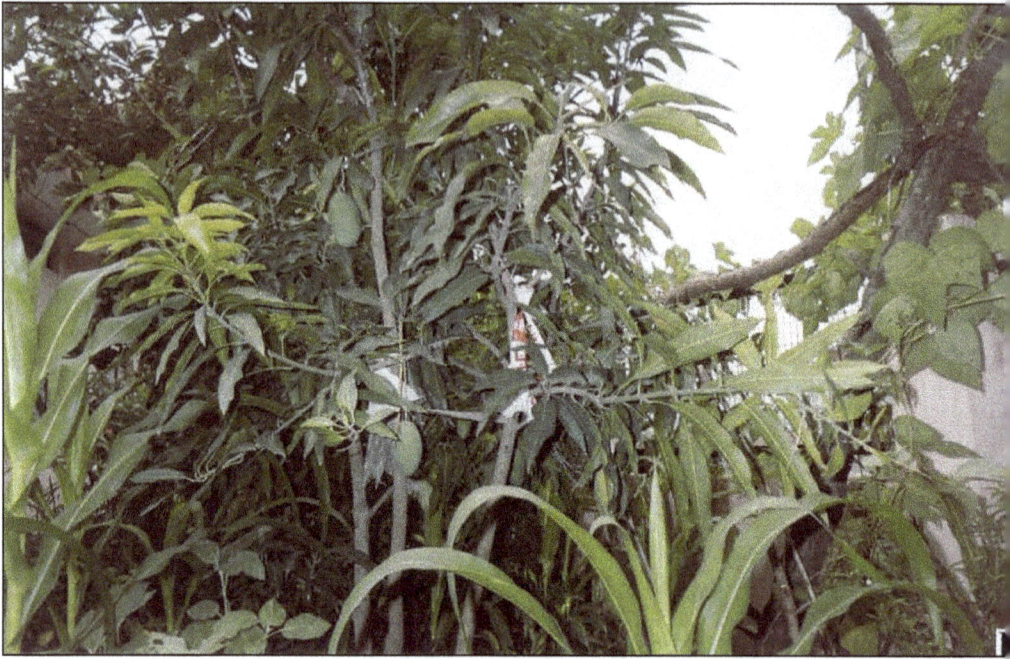

Plate 12: *Mangifera indica* Linn.

Plate 13: *Piper nigrum* Linn.

Plate 14: Flowering Branch of
***Punica granatum* Linn.**

Plate 15: *Solanum melongena* Linn.

Plate 16: *Syzygium aromaticum* Merr. & Perry

Plate 17: Seeds of *Trigonella foenum-graecum* Linn.

Plate 18: Tree of *Syzygium cumini* (Linn.) Skeels.

Plate 19: Flowers of *Trigonella foenum-graecum* Linn.

www.ingramcontent.com/pod-product-compliance
Lightning Source LLC
Chambersburg PA
CBHW050513190326
41458CB00005B/1519